THE ENIGMA OF CRANIAL DEFORMATION

Elongated Skulls of the Ancients

David Hatcher Childress
&
Brien Foerster

Other Books by David Hatcher Childress:

THE MYSTERY OF THE OLMECS
PIRATES AND THE LOST TEMPLAR FLEET
TECHNOLOGY OF THE GODS
A HITCHHIKER'S GUIDE TO ARMAGEDDON
LOST CONTINENTS & THE HOLLOW EARTH
ATLANTIS & THE POWER SYSTEM OF THE GODS
THE FANTASTIC INVENTIONS OF NIKOLA TESLA
LOST CITIES OF NORTH & CENTRAL AMERICA
LOST CITIES OF CHINA, CENTRAL ASIA & INDIA
LOST CITIES & ANCIENT MYSTERIES OF AFRICA & ARABIA
LOST CITIES & ANCIENT MYSTERIES OF SOUTH AMERICA
LOST CITIES OF ANCIENT LEMURIA & THE PACIFIC
LOST CITIES OF ATLANTIS, ANCIENT EUROPE & THE MEDITERRANEAN
LOST CITIES & ANCIENT MYSTERIES OF THE SOUTHWEST

Other Books by Brien Foerster:

A BRIEF HISTORY OF THE INCAS
MACHU PICCHU VIRTUAL GUIDE & SECRETS REVEALED
THE ENIGMA OF TIWANAKU AND PUMA PUNKU
GRINGO GUIDE MACHU PICCHU AND CUSCO
INCA BEFORE THE CONQUEST
BEYOND MACHU PICCHU:
THE OTHER MEGALITHIC MONUMENTS OF ANCIENT PERU

THE ENIGMA OF CRANIAL DEFORMATION

Adventures Unlimited Press

The Enigma of Cranial Deformation

ISBN 10: 1-935487-76-0
ISBN 13: 978-1-935487-76-0

Published by:
Adventures Unlimited Press
One Adventure Place
Kempton, Illinois 60946 USA
auphq@frontiernet.net

www.adventuresunlimitedpress.com

10 9 8 7 6 5 4 3 2 1

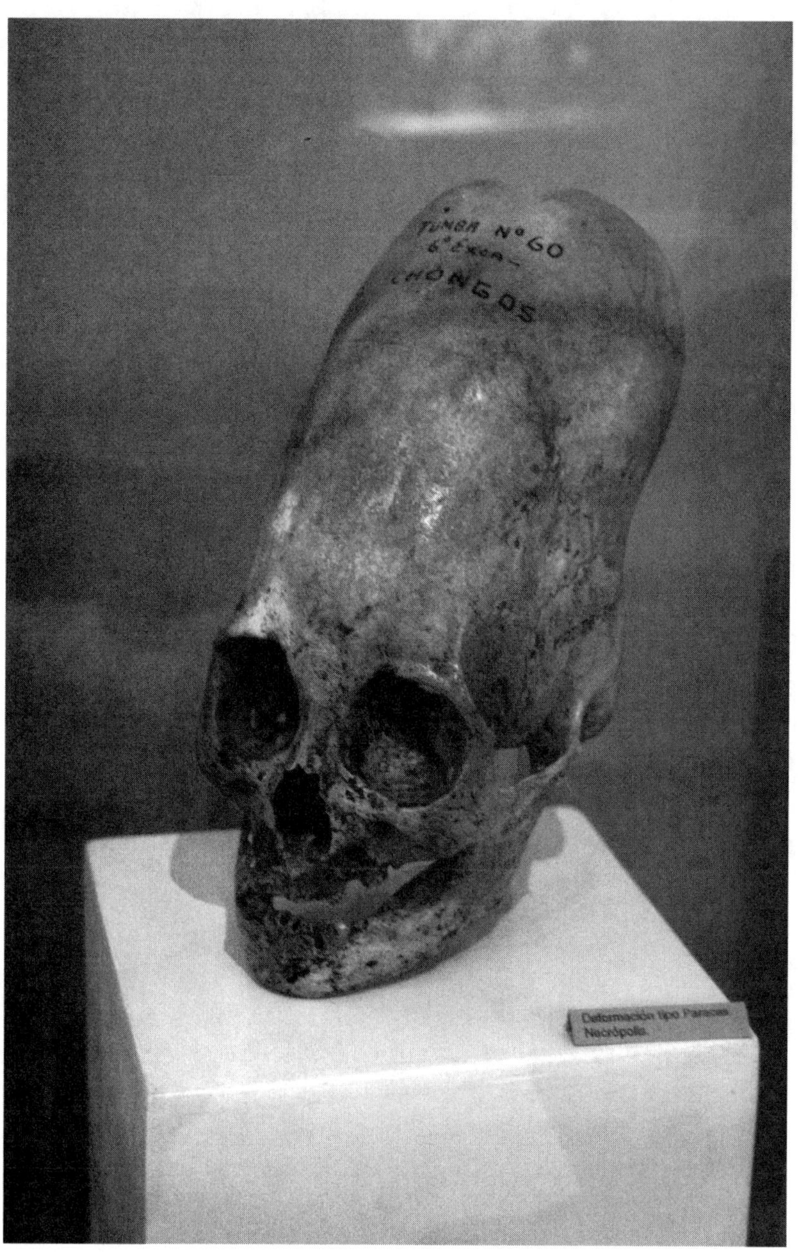

One of the extremely elongated skulls from the Chongos site near Paracas.

Contact David Hatcher Childress at:
www.wexclub.com

Contact Brien Foerster at:
www.HiddenIncaTours.com

For more copies of this book
please go to:
www.adventuresunlimitedpress.com

THE ENIGMA OF CRANIAL DEFORMATION

Elongated Skulls of the Ancients

David Hatcher Childress
&
Brien Foerster

TABLE OF CONTENTS

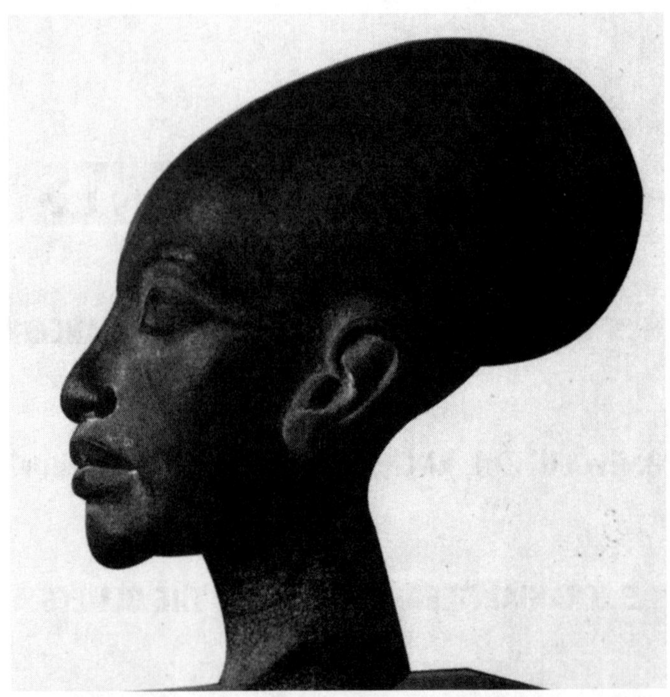

A bust of the Atonist princess Meketaton now at the Cairo Museum.

One of the extremely elongated skulls from Paracas, Peru.

A set of extremely elongated skulls from Paracas, Peru.

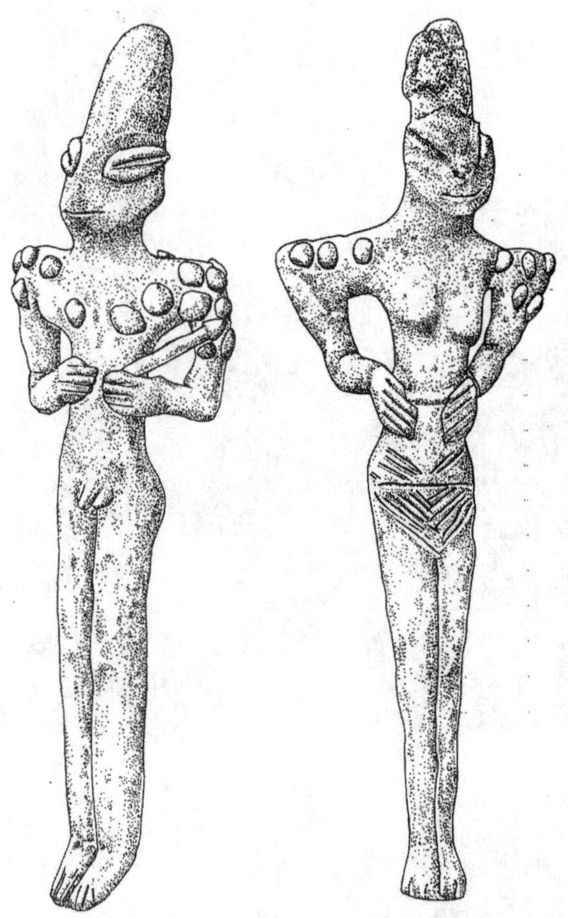

Two Sumerian clay figurines from the Ubaid Period (5000 BC) showing elongated heads. The male on the left is from Eridu and the female on the right is from the ancient city of Ur.

CHAPTER 1

MYSTERIOUS ELONGATED SKULLS OF THE ANCIENTS

The human race is over a million years old,
But it never seems to act its age.
—Professor Hiernonymus

The Mystery of the Strange Coneheads

Many of you may remember the old Saturday Night Live routine with Dan Akroyd and gang as the strange family of "Coneheads" (later made into a feature movie). This bizarre family of comical weirdos had long, bald heads that were about twice as long as a normal human head—heads which came to a sharp point at the top, hence the name "coneheads."

As funny and strange as this may appear to television viewers, such "coneheads" did, and do, exist! In fact, the remains of "coneheads" of various shapes and sizes have been found all over the world, from Peru and Mexico and the Pacific Northwest to ancient Egypt and China and beyond.

Museums in South America and elsewhere have occasionally exhibited an unusual elongated skull, but they are not particularly common in museums around the world. At the local museums along the southern desert coast of Peru, particularly the anthropological museum in Ica, were a wide variety of elongated skulls and even complete mummies of these people, sometimes with red hair. Similarly, statues with elongated skulls of the Atonists can be seen at the Cairo Archeological Museum. Other museums, in Malta, Turkey, Iraq, Korea, Bolivia, Mexico and elsewhere have skulls with some sort of cranial deformation on exhibit. Many of them, however, are not on display. In some museums, such as the Archeological Museum at Tiwanaku in Bolivia, photography is not allowed (although sneaky tourists have managed to take a few photos).

To the uninitiated, the deformed and extended skulls are rather shocking. They come in all shapes and sizes, some extremely elongated, and some rather squarish, instead of pointed. Just exactly who these people were, why they reshaped heads, and how they created the odd modifications are something of a mystery. The skull elongations are the most prevalent; perhaps we should look into some of the various explanations for these coneheads.

This book will take us to Peru, Bolivia, Egypt, Malta, China and other places in search of strange elongated skulls and other cranial deformation. The puzzle of why diverse ancient people—

A poster for the 1993 *Coneheads* movie starring Dan Akroyd.

even on remote Pacific Islands—would use headbinding to create elongated heads is mystifying. Where did they even get this idea? Did some people naturally look this way—with long narrow heads? Were they some sort of elite race that roamed the entire planet? Why do anthropologists rarely talk about cranial deformation and know so little about it?

Types of Cranial Deformation

There are two main types of cranial deformation with a number of sub-types, as we shall see. Mainly skulls are deformed by either being elongated or alternately flattened and made wider. Says the *Encyclopedia Britannica*:

> Modifications of the head have included alterations of the skull, teeth, lips, tongue, nose, or ears. Deformation of the skull is the best-documented form, largely because archaeological skeletal remains clearly show its presence. Tabular deformations are produced by constant pressure of small boards or other flattened surfaces against the infant's head (*see* head flattening). Annular deformations are produced by a constricting band; each kind is subdivided according to the resulting head shape, which is often strikingly different from the unmodified skull. Cases of cranial modification are known from all continents except Australia and Oceania, although it was rather rare in Africa south of the Sahara and apparently absent from South India.

Let us look at information on cranial deformation from the online encyclopedia, Wikipedia:

> Artificial cranial deformation, head flattening, or head binding is a form of permanent body alteration in which the skull of a human being is intentionally deformed. It is done by distorting the normal growth of a child's skull by applying force. Flat shapes, elongated ones (produced by binding between two pieces of wood), rounded ones (binding in cloth) and conical ones are among those chosen.

It is typically carried out on an infant, as the skull is most pliable at this time. In a typical case, headbinding begins approximately a month after birth and continues for about six months.

Intentional head molding producing extreme cranial deformations was once commonly practiced in a number of cultures widely separated geographically and chronologically, and so was probably independently invented more than once. It still occurs today in a few groups, like the Vanuatu. Early examples of intentional human cranial deformation predate written history and date back to 45,000 BC in Neanderthal skulls, and to the Proto-Neolithic Homo sapiens component (12th millennium BC) from Shanidar Cave in Iraq. It occurred among Neolithic peoples in SW Asia.

The earliest written record of cranial deformation dates to 400 BC in Hippocrates' description of the Macrocephali or Long-heads, who were named for their practice of cranial modification. The practice was also known among the Australian Aborigines, Maya, and certain tribes of North American natives, most notably the Chinookan tribes of the Northwest and the Choctaw of the Southeast.

In the Old World, Huns and Alans are also known to have practiced similar cranial deformation. In Late Antiquity (AD 300-600), the East Germanic tribes who were ruled by the Huns, adopted this custom (Gepids, Ostrogoths, Heruli, Rugii and Burgundians). In western Germanic tribes, artificial skull deformations have rarely been found...

Friedrich Ratzel in *The History of Mankind* reported in 1896 that deformation of the skull, both by flattening it behind and elongating it towards the vertex, was found in isolated instances in Tahiti, Samoa, Hawaii, and the Paumotu group and occurring most frequently on Mallicollo in the New Hebrides, where the skull was squeezed extraordinarily flat.

As to the methods and reasons for cranial deformation and headbinding, both of which are poorly studied and continue to baffle scholars, Wikipedia says the following:

> Deformation usually begins just after birth for the next couple of years until the desired shape has been reached or the child rejects the apparatus (Dingwall, 1931; Trinkaus, 1982; Anton and Weinstein, 1999). There is no established classification system of cranial deformations. Many scientists have developed their own classification systems, but none have agreed on a single classification for all forms that are seen (Hoshower et al., 1995).
>
> In Europe and Asia three main types of artificial cranial deformation have been defined by E.V. Zhirov (1941, p. 82): Round, Fronto-occipital, Sagittal.
>
> Cranial deformation was probably performed to signify group affiliation, or to demonstrate social status. This may have played a key role in Egyptian and Mayan societies. Queen Nefertiti is often depicted with what may be an elongated skull, as is King Tutankhamen. It could be aimed at creating a skull shape which is aesthetically more pleasing or associated with desirable attributes. For example, in the Nahai-speaking area of Tomman Island and the south south-western Malakulan, a person with an elongated head is thought to be more intelligent, of higher status, and closer to the world of the spirits.

Unfortunately, Wikipedia is not a very good source on cranial deformation, though little information can be found on the subject and books on the subject are noticeably absent. For instance, the Wikipedia article says that head binding "continues for about six months" and then later says "...usually begins just after birth for the next couple of years..." Although the Wikipedia article does not mention it, the scientific term for people with elongated or flattened cranium is dolichocephaloid or brachycephaloid. We can call them cranially deformed, coneheads or flatheads, but science will call them by the difficult to pronounce word dolichocephaloid.

Peruvian museum illustration of typical headbinding.

Australian researcher Karen Mutton[2] defines dolichocephaloids in her book *Scattered Skeletons in Our Closet*:

> A dolichocephaloid skull is by definition long and narrow as opposed to a brachycephaloid skull which is broader and rounder. Historically, dolichocephalic skulls generally belong to Caucasian skeletons and brachiocephalic skulls belong to Asiatic skeletons although there are variations. In recent years extremely dolichocephalic, conical skulls have been discovered in South American museums which defy explanation. It is well known that many ancient cultures practiced head binding which distorted skull shapes but extreme dolichocephalism has rarely been studied by anthropologists.
>
> Cone headed skulls were first mentioned [in modern times] in 1851 in the book 'Peruvian Antiquities' by Mariano Rivero and John James von Tschudi. Dr. Tschudi, with credentials in philosophy, medicine and surgery, described dolichocephalism in two distinct Peruvian races that existed before the Incas, the Aymares and Huancas.

The Huancas had the most pronounced dolichocephalic traits although Tschudi had little historical data on them. The Aymaraes had intermediate dolichocephaly.

Even at that time scientists were proclaiming that the skulls had been artificially elongated by head binding, a claim currently in favor for the Australian Kow Swamp skeletons. However Dr. Tschudi commented, "Two crania (both of children scarcely a years old), had in all respects, the same form as those of adults. We ourselves have observed the same fact in many mummies of children of tender age... The same formation of the head presents itself in children yet unborn, and of this truth we have had convincing proof in sight of a foetus enclosed in the womb of a mummy of a pregnant woman...aged 7 months!"

Researcher Lumir Janku has studied many of the anomalous skulls from the Paracas region and divided them into 3 types; premodern, Conehead 1,2 and 3. The 'premodern' skull has some pre Neanderthal features such as pronounced brow ridges, robust lower jaw and occipital ridge on the bottom back of the skull. Its massive cranial arch, to Janku, suggests "that the skull belongs to a representative of an unknown premodern or humanoid type." His illustration indicates that the brain of the 'premodern' was of a similar capacity to modern humans, despite its extreme elongation.

The Cone heads, as evidenced from three different specimens, C1, 2 and 3 may have developed from the 'premodern' type but reveal a much larger brain.

Dolichocephalic skulls have also been discovered in some of the earliest Old World cultures of Malta and Iraq. Janku wrote, "The enormity of the cranial vault is obvious from all three pictures. By interpolation, we can estimate the minimum cranial capacity at 2200 ccm, but the value can be as high as 2500 ccm." This can be compared to modern humans whose brain capacity is, on average about 1450 ccm.

He believes it is a distinct branch of the genus Homo,

if not an entirely different species.

Different Theories on Who These People Collectively Were

There are a number of different theories on the people with elongated heads. It is generally assumed that most of the coneheads were originally normal human beings with normal heads. However, there is also the popular theory that some people naturally had elongated heads and just looked this way. Others then decided to imitate them and artificially create the unusual and sometimes startling elongated heads (as well as other types). Let us take a look at some of the different theories.

The Atlantis Theory

One theory on these cranially deformed individuals is that they are remnants of the citizens of the lost continent of Atlantis. In this theory, Atlanteans, for reasons unknown, liked to have long, conical shaped skulls. According to this theory it was popular to have the head shaped in infancy to double the size of the skull and increase brain capacity.

It is thought that surviving Atlanteans, in their worldwide journeys, impressed other cultures with their high level of civilization and knowledge of all things psychic and scientific. Inhabitants of the colonies they founded, such as those in Mexico, Peru and Egypt, began to imitate the Atlanteans and their unusual customs such as head binding. A similar theory says that survivors of the lost continent of Mu in the Pacific to also disseminate coneheads to South America and Mexico.

The Nephilim-Watchers Theory

A related theory involves the "Watchers" mentioned in *The Book of Enoch*. They appear to have had elongated heads, and British researcher Andrew Collins believes that they originally came from Kurdistan in northern Iraq, but also maintained communities in the mountains of Lebanon and other areas. These people were said to have tremendous knowledge and even the power of flight. They allegedly went all over the world in boats and airships.

The large boats of the Watchers need no real explanation—we see them in the large Phoenician, Arabian and Chinese vessels known to exist. The airships of the watchers are a stranger topic. They are the same as the famous vimanas of the epic Hindu texts such as the Ramayana—airships in antiquity that could reach distant lands and are still familiar today to every Hindu and Buddhist.

This would help to explain why cranial deformation is so widespread. It can be found in ancient Sumeria, among the "Watchers" of Kurdistan, and among certain of the Egyptian royalty. This also fits in with the theories that the Olmecs and the Tiahuanaco culture of Peru (and Bolivia) that practiced headbinding were aligned with the remnants of the Atlantean civilization, sometimes called the Atlantean League.

This theory also supports the theory of diffusionism and ancient seafaring that brought cultural diffusion across both the Atlantic and Pacific Oceans. Such unusual—yet widespread—customs as cranial deformation, turban wearing, jade worship, making keystone cuts in megalithic masonry and trepanning can best be understood as having been transferred from one culture to another by ancient contact between widely separated cultures.

The Nephilim-Extraterrestrial Theory

This theory largely maintains that giant "space aliens" came to our planet and helped shape ancient civilizations. These extraterrestrials had elongated heads and oriental features (such as narrow, slanting eyes, often called "coffeebean eyes") but did not artificially create elongated skulls. They looked like this naturally. They were seven to nine feet tall or more.

Statues and other figurines found at ancient Sumerian sites in Iraq indicate that some of the inhabitants of the area had elongated heads and the puffy, slit-like "coffeebean" eyes. Such figurines are thought to give the impression of a reptilian countenance, and are today called by archeologists "serpent priests" or "lizard" figurines. Were they humans, or extraterrestrials?

In the extraterrestrial theory, the ETs were the overlords of the primitive Earth populations (which were possibly genetically engineered by them) and were highly admired by the humans who were their subjects. As these extraterrestrials returned to their own planet, the humans left in charge decided to emulate the aliens who had elongated heads and looked distinctly different from normal humans; they instituted the practice of skull deformation in order to try to look like the extraterrestrial masters who had once ruled them.

The Mainstream

Mainstream science has no real theory on why so many diverse cultures took up the practice of head binding. Most concur that it was elitist practice to set some apart from the masses. Researchers have focused on the procedures used to produce the strange deformities. It seems that the skulls were created while the humans were still infants. For unknown reasons, infants would have their heads bound with pieces of wood and rope, or some sort of constrictive cloth, that forced the heads to grow in an elongated and unnatural way.

While the plates in a baby's skull have not yet fused together, the skull is bound with materials that are adjusted and tightened as time goes on. After several years of this skull binding, the child's head is permanently growing in an elongated fashion and once the child becomes several years old the skull has formed in an elongated fashion that no longer requires any binding.

From this point on, the person's head will continue grow in an elongated fashion until growth stops as a late teenager. Now the person is an otherwise normal adult human, except that he has an elongated skull and sometimes nearly double the brain capacity of other humans who have not had their heads bound. As mentioned in the Wikipedia article above, this was assumed to be "practiced in a number of cultures widely separated geographically and chronologically, and so was probably independently invented more than once."

Trepanning and Other Skull Surgery

Like cranial deformation, there are the curious holes often cut into skulls worldwide. The process of drilling or sawing holes in the skull is called "trepanning" and evidence of this practice has been found on elongated skulls as well as on normal skulls. Trepanning can be a round hole in the skull or a square hole made with four sawing cuts into the skull. That people survived this ancient "brain surgery" is evident from skulls that have been found that exhibit

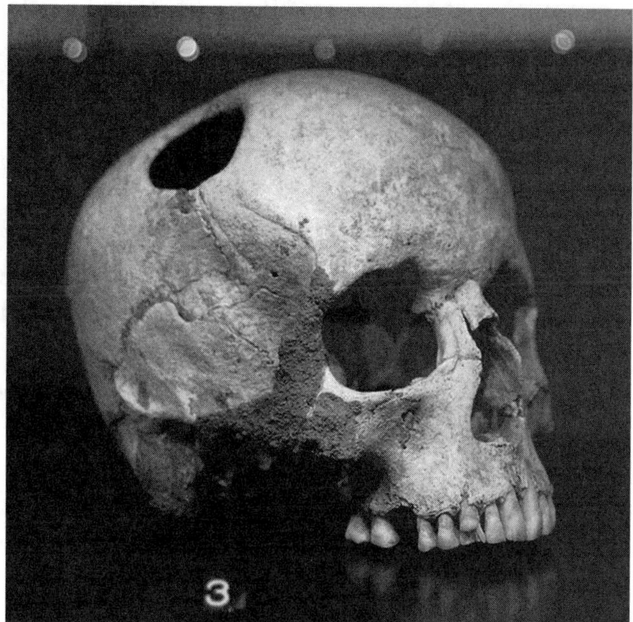

A trepanned skull from Lausanne, France, circa 3500 BC.

further calcium growth around the holes.

Again, the question is why?

Many deformed crania found in South America and other areas have been found to have been trepanned. In order to get a basic understanding of trepanning, let us look at Webster's Online Dictionary. According to Webster's the definition is:

> Trepanation (also known as trepanning, trephination, trephining or burr hole) is surgery in which a hole is drilled or scraped into the skull, thus exposing the dura mater in order to treat health problems related to intracranial diseases, though in the modern era it is used only to treat epidural and subdural hematomas, as an extreme body modification, and for surgical access for certain other neurosurgical procedures, such as intracranial pressure monitoring.
>
> Trepanation was carried out for both medical reasons and mystical practices for a long time: evidence of trepanation has been found in prehistoric human remains from Neolithic times onwards, per cave paintings indicating

that people believed the practice would cure epileptic seizures, migraines, and mental disorders. Furthermore, Hippocrates gave specific directions on the procedure from its evolution through the Greek age.

The modern medical procedure of corneal transplant surgery uses a technique known as trepanning or trephining, however the operation is conducted on the eye (not the skull), with an instrument called a trephine.

Trepanation is perhaps the oldest surgical procedure for which there is evidence, and in some areas may have been quite widespread. Out of 120 prehistoric skulls found at one burial site in France dated to 6500 BC, 40 had trepanation holes. Surprisingly, many prehistoric and premodern patients had signs of their skull structure healing; suggesting that many of those that proceeded with the surgery survived their operation.

Trepanation was also practiced in the classical and Renaissance periods. Hippocrates gave specific directions on the procedure from its evolution through the Greek age, and Galen elaborates on the procedure, too. Doctors in ancient Egypt used the scrapings of the skull to create love potions and other concoctions.

During the Middle Ages and the Renaissance, trepanation was practiced as a cure for various ailments, including seizures and skull fractures. The surgeons who performed these trepanations were probably highly skilled because the survival rate of the operations was high and the infection rate was low.

An Incan trepanned skull from Cuzco.

Though trepanning has been done to perfectly normal skulls, many of the elongated skulls have also been trepanned. Many of the skulls seen in South America have even had gold plates inserted in the trepanned hole. Because we can see that the bone material has grown over the gold plate, or otherwise healed, it is obvious that many of the patients went on to live

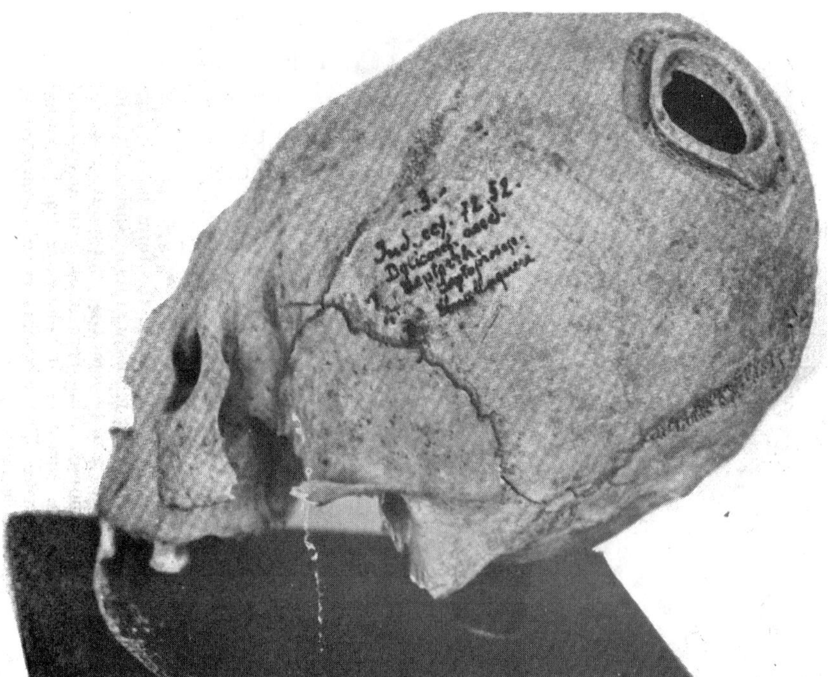

Rare double trepanning on an elongated skull from Tiwanaku, now at the Archeologiy Museum in La Paz, Boliva.

for many years after their trepanning.

Why were they trepanned? We can only assume that they had headaches or voices in their head that needed expulsion. The popular mainstream theory is that trepanning was a type of brain surgery for ancient people with mental problems. These cultures believed that a person with a mental problem such as schizophrenia or hearing voices was possessed by evil spirits or the like. By cutting a hole in the mentally ill person's head they could free the evil spirits that were causing the problem and the "patient" would hopefully get better. Did people with elongated skulls suffer from headaches and voices more than people with normal crania? We will probably never know.

On the other hand, there is the curious notion that a hole drilled in one's skull can actually enhance one's psychic abilities. As far as I know, this was first put forward into popular literature by the British author T. Lobsang Rampa in his 1960 book *The Third Eye*. In this book, the supposed true account of the author's life as a doctor-in-training at a Tibetan monastery in the 1920s, Rampa

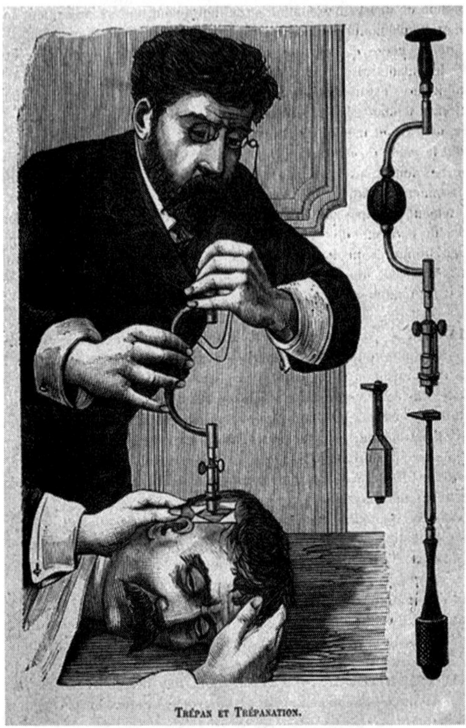

Old print depicting a trepanning operation.

reports that he had a hole drilled in his forehead while a young man to release his latent psychic powers.

A hole was made with a crude hand drill that cored out a two-inch hole in his forehead. The hole was then plugged with a piece of special wood, and an herbal salve was placed over the entire area. Rampa describes it as a traumatic experience, but afterward he claims that he could more easily see patients' etheric auras, etc. While it is doubtful that *The Third Eye* is total non-fiction as the author maintains, it is a fascinating story nonetheless.

Such surgeries were performed as well in the 1960s in Britain and Holland by a small group of eccentrics who believed that by opening holes in the tops of their heads they would be permanently "high" and, theoretically, have enhanced psychic powers. British author John Michell chronicles some of these cases in his 1984 book *Eccentric Lives and Peculiar Notions*[8] in a chapter entitled "The People with Holes in their Heads."

Michell says that the founder of the modern trepanation movement is a Dr. Bart Hughes from the Netherlands. Hughes made a discovery in 1962 that one's state and degree of consciousness are related to the volume of blood in the brain. According to his theory of evolution, the adoption of an upright stance brought certain benefits to the human race, but it caused the brain to be partially starved of oxygen because of the effects of gravity and the difficulty of getting enough blood to the head. One can redress the balance by a number of methods, such as standing on one's

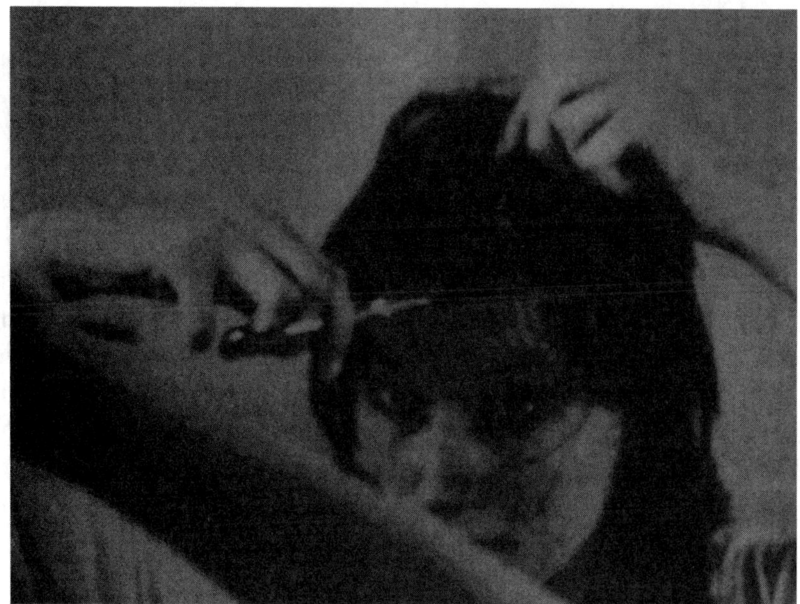

Amanda Feilding preparing for her voluntary trepanning.

head, jumping from a hot bath into a cold one, or the use of drugs; but the wider consciousness thus obtained is only temporary. Bart Hughes believed that a hole in his head would permanently increase blood to his brain and help counter the effects of gravity.

In 1965 a Londoner named Joseph Mellen met Bart Hughes on the Spanish resort island of Ibiza and quickly became Hughes' leading (perhaps only) disciple. He too drilled a hole in his head and wrote an obscure book a few years later titled *Bore Hole*. Later, Mellen met another eccentric Londoner named Amanda Feilding whom he began living with. She was impressed by Mellen and the hole in his head and decided to drill a hole in her own head. Feilding actually made a short film of her self-inflicted operation entitled *Heartbeat in the Brain*. In the film Feilding shaves her head and then drills a hole with an auger, blood spurting out as she penetrates the skull (this film is now lost).

Making the Head So Much Larger

On the practical side, one must acknowledge that the elongated deformation of skulls does enlarge the size of the cranium and appears to allow the brain to expand to a much larger size than would otherwise be normal. Does this mean that people with dolichocephalous crania had bigger brains and were therefore smarter or more psychic than humans with normal crania? It is an interesting thought considering that we are often told by researchers that we don't use the full capacity of our brains anyway. Someone with a larger skull and larger brain may be making more use of his brain and some therefore have an advantage over people with normal size brains.

While it is assumed that headbinding will allow the cranium to enlarge and stretch, creating a larger volume in a skull than would be in a normal person, not all researchers believe that this is possible. Some researchers maintain that headbinding would not enlarge the volume in a person's skull as the amount of skeletal material would not increase, only the shape of the skull would change. Until more research is done on this subject, we will not have a definite answer on whether headbinding can double the volume of a crania.

So, we can surmise that the ancients practiced cranial deformation for a number of reasons. They wanted longer skulls to look different from normal people. They wanted to enlarge their brains, perhaps for psychic reasons. They wanted to imitate some sort of elite that were seen as superior to the normal peasants of the various regions where cranial deformation was practiced.

What is particularly interesting, and unexplained by the standard writers on the subject, is how and why cranial deformation was practiced in nearly every continent and includes groups of people on remote islands and remote jungles. How did this curious practice spread all over the world?

Old print depicting a trepanning.

Hieronymus Bosch painting of a trepanning operation.

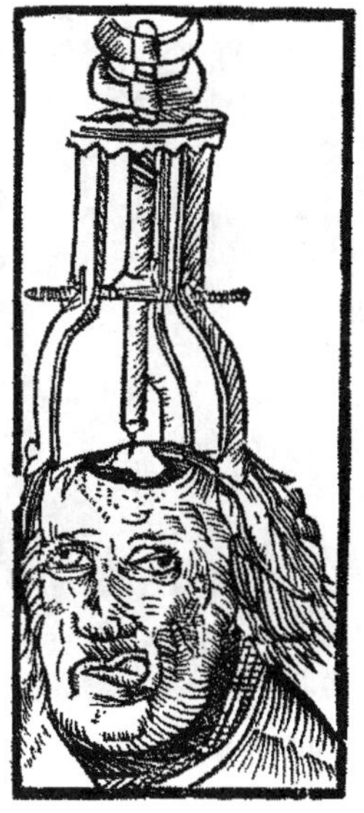

Dutch artist Peter Treveris' woodcuts on trepanning.

A trepanned skull from the Archeological Museum in Ica, Peru.

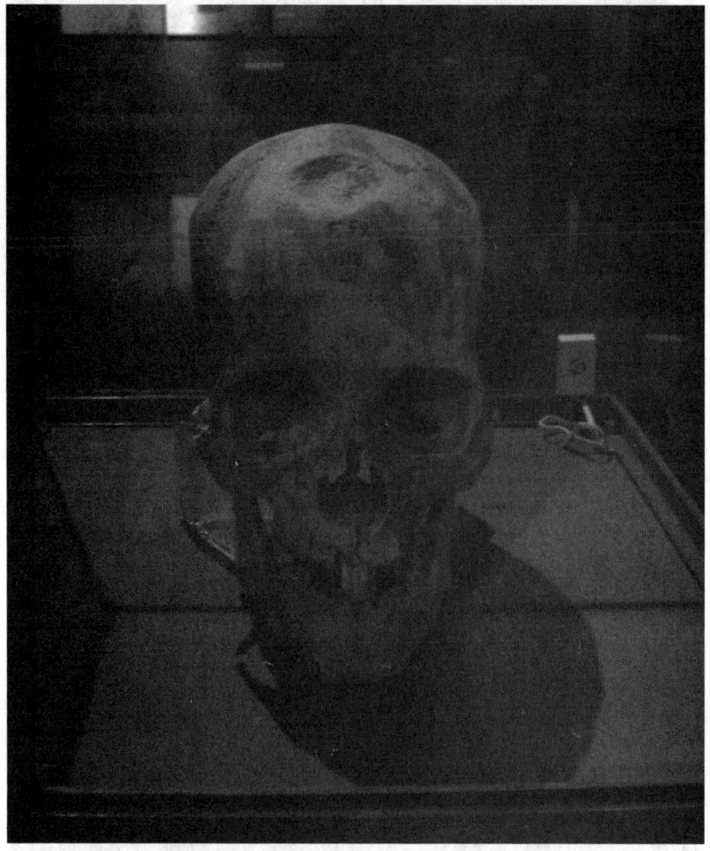

A trepanned skull with skin covering the hole from Luxor, Egypt.

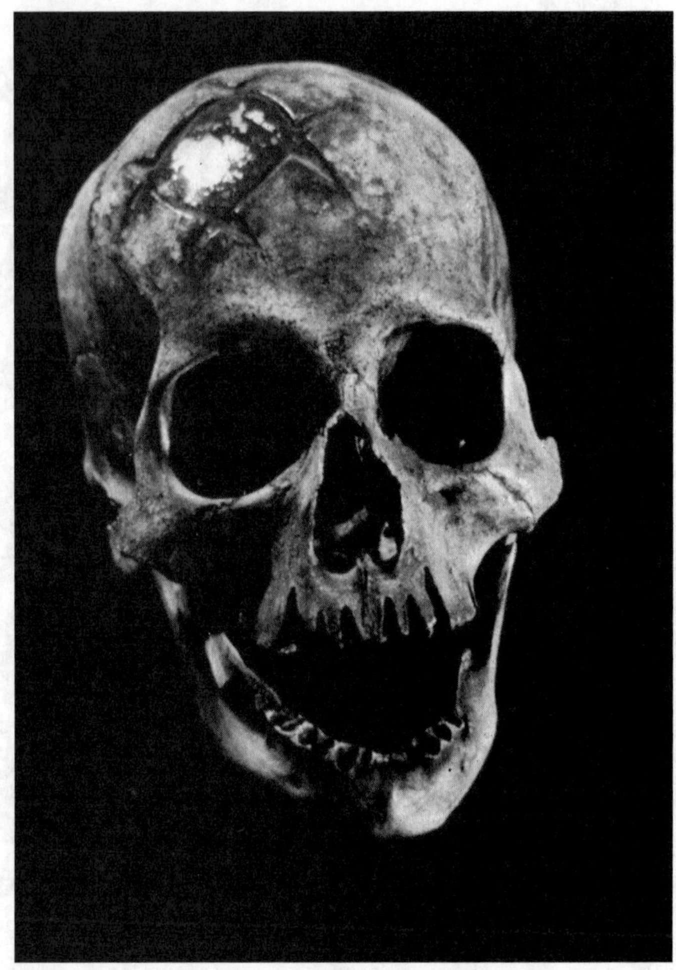

A trepanned skull with a gold plate from the Gold Museum, Lima, Peru.

CHAPTER 2

EVOLUTION, ANCIENT MAN AND THE CRANIUM

Evolution is like walking on a rolling barrel:
The walker isn't so much interested in where
the barrel is going
as in keeping on top of it.
—Robert Frost

Where have we come from? Did we once have an ancestor who had a naturally elongated head? Or did ancient man, through random experimentation, create the coneheads of the ancient world? In order to understand the elongated skulls of ancient times, we need to understand the concepts of evolution and "intelligent evolution" or a guided manipulation of humanity. Was mankind manipulated by extraterrestrials, or "angels"? Did a separate race with naturally elongated skulls evolve at the same time as other humans? Did the practice of cranial deformation "evolve" over time with our distant ancestors?

The history of humanity has been well documented by archaeologists and anthropologists, and fits into a series of time frames and evolutionary advances bringing us from ancestral ape-like forms to the *Homo sapiens* which make up the population of the present day earth—or so they would like us to believe. The truth may be far more complex and intriguing than a simple model of "from ape to human," as put forth in the Darwinian model, created in the 19th century. This model could not take into consideration a lot of research and discoveries that occurred after Charles Darwin and his contemporaries published, and in fact lived.

The term "human" in the context of human evolution refers to the genus *Homo*, but studies of human evolution usually include other hominids, such as the Australopithecines, from which the genus *Homo* had diverged by about 2.3 to 2.4 million years ago in Africa.[31] Many scientists have estimated that humans branched off from a common ancestor with chimpanzees about five to seven million years ago.

Several species and subspecies of *Homo* evolved and are now extinct; these are thought of as either side branches of the overall evolutionary line, or possibly direct or indirect ancestors of modern humans which perished due to environmental changes, and/or competition pressures. They include *Homo erectus*, which inhabited Asia, and *Homo sapiens neanderthalensis*, which inhabited Europe. So called "archaic" *Homo sapiens*, our supposed direct ancestors, are believed to have evolved, initially, between 400,000 and 250,000 years ago.[32]

The dominant view among scientists concerning the origin of anatomically modern humans is the "Out of Africa" or "Recent African Origin" hypothesis, which argues that *Homo sapiens* arose in Africa, exclusively, and migrated out of the continent around 50,000 to 100,000 years ago, replacing populations of *Homo erectus* in Asia and *Homo neanderthalensis* in Europe.[33] Scientists that are supporting the alternative "Multiregional Hypothesis" argue that *Homo sapiens* evolved as geographically separate but interbreeding populations stemming from a worldwide migration of *Homo erectus* out of Africa nearly 2.5 million years ago.

The word *Homo*, the name of the biological genus to which humans belong, is Latin for "human." It was chosen originally by Carolus Linnaeus (1707-1778) in his famous classification system. The word "human" is from the Latin *humanus*, the adjectival form of *homo*.[34] Linnaeus and other scientists of his time considered the great apes to be the closest relatives of humans due to morphological and anatomical similarities. The possibility of linking humans with earlier apes by descent only became clear after 1859 with the publication of Charles Darwin's *On the Origin of Species*. This treatise argued for the idea of the evolution of new species from earlier ones. Darwin's book did not address the question of human evolution, however, claiming only that, "Light will be thrown on the origin of man and his history."[35]

The first debates about the nature of human evolution arose

between Thomas Huxley, a colleague of Darwin nicknamed "Darwin's Bulldog" and his respected contemporary Richard Owen. The latter agreed with Darwin that evolution occurred, but thought it was more complex than outlined in Darwin's famous *Origin*. Owen was also, by the way, the driving force behind the establishment of the British Museum in London in 1881.

An 1871 editorial cartoon depicting Darwin as a monkey.

Huxley argued for human evolution from apes by illustrating many of the similarities and differences between humans and apes, and did so particularly in his 1863 book entitled *Evidence as to Man's Place in Nature*. However, many of Darwin's early supporters, such as Alfred Russell Wallace and Charles Lyell did not agree that the origin of the mental capacities and the moral sensibilities of humans could be explained by Darwin's theory of natural selection. In response to this, Darwin applied the theory of evolution and sexual selection to humans when he published his next book, *The Descent of Man*[35] in 1871.

A major problem in finding a smooth, flowing sequence of evolution from one species to the next was the lack of fossil intermediaries, as in "missing links." It was only in the 1920s that the first of such fossils was discovered in Africa. In 1925, Raymond Dart described *Australopithecus africanus,* which was thought by some researchers to be the first true human ancestor. The specimen described was named the Taung Child, an Australopithecine infant discovered in a cave. The child's remains included a remarkably well-preserved tiny skull, and though the brain cavity was relatively small (410 cm^3), its shape was rounded, unlike that of chimpanzees and gorillas, and more like that of a modern human skull. Also, the found specimen showed short canine teeth, and the position of the foramen magnum (the hole in the skull where the spine enters) was evidence of bipedal locomotion. All of these traits convinced Dart that the Taung baby was a bipedal human ancestor, a transitional form between apes and humans.[36]

Theoretically, the next in line in evolutionary history, to the supposed first "human like" great apes, *Homo habilis,* lived from about 2.4 to 1.4 million years ago. It evolved in South and East Africa in the late Pliocene or early Pleistocene, 2.5 to 2 million years ago, when it supposedly diverged from the Australopithecines. *Homo habilis* had smaller molars and larger brains than the Australopithecines, and made tools from stone and perhaps animal bones. One of the first known hominids, it was nicknamed "handy man" by its discoverer, the famous Louis Leakey, due to its association with stone tools.[37] Some scientists have proposed moving this species out of *Homo* and into *Australopithecus* because the morphology of its skeleton was more adapted to living in trees than moving on two legs like *Homo sapiens*.

The following are proposed species names for fossils from about 1.9 to 1.6 million years ago, the relation of which with *Homo habilis* is not yet clear.

1) *Homo rudolfensis* refers to a single, incomplete skull from Kenya. Scientists have suggested that this was another *Homo habilis*, but this has not yet been confirmed.

2) *Homo georgicus*, from the country of Georgia, may be an intermediate form between *Homo habilis* and *Homo erectus*, or even a sub-species of *Homo erectus*.

In the early Pleistocene, 1.5 to 1 million years ago, in Africa, Asia and Europe, some populations of *Homo habilis* are thought to have evolved larger brains and made more elaborate stone tools; these differences and others were sufficient for anthropologists to classify them as a new species, *Homo erectus*. In addition, to this *Homo erectus* was the first human ancestor to walk truly upright. This attribute was made possible by the evolution of locking knees and a different location of the foramen magnum. They may have also used fire to cook their meat.

The first fossils of *Homo erectus* were discovered by Dutch physician Eugene Dubois in 1891 on the Indonesian island of Java. He originally gave the skeletal material the name *Pithecanthropus erectus* based on its morphology, which he considered to be intermediate between that of apes and humans.[38] A famous example of *Homo erectus* is Peking Man, while others were found in Asia (notably in Indonesia), Africa and Europe.

There are also proposed species that may be intermediate between *Homo erectus* and *Homo heidelbergensis* (which is thought to be an ancestor of modern humans—see below):

1) *Homo antecessor* is known from fossils from Spain and England that are dated 1.2 million to 500,000 years ago.

2) *Homo cepranensis* refers to a single skull cap from Italy, estimated to be about 800,000 years old.

Homo heidelbergensis (Heidelberg Man) lived from about 800,000 to about 300,000 years ago. *H. neanderthalensis*, commonly known as Neanderthals, lived from 400,000 to about 30,000 years ago. Evidence from sequencing mitochondrial DNA indicated that no significant gene flow occurred between *H. neanderthalensis* and *H. sapiens*, and, therefore, the two were separate species that shared a common ancestor about 660,000 years ago. In 1997, Mark Stoneking stated: "These results [based on mitochondrial DNA extracted from Neanderthal bone] indicate that Neanderthals did not contribute mitochondrial DNA to modern humans… Neanderthals are not our ancestors."[39] Subsequent investigation of a second source of Neanderthal DNA supported these findings.

However, a 2010 sequencing of samples of the Neanderthal genome indicated that Neanderthals did indeed interbreed with *H. sapiens* circa 75,000 BC (after

A reconstruction of a Neanderthal.

H. sapiens moved out from Africa, but before they separated into Europe, the Middle East and Asia). Nearly all modern humans, at least non-African humans, have 1% to 4% of their DNA derived from Neanderthal DNA. Supporters of the Multiregional Hypothesis point to recent studies indicating the presence of non-African nuclear DNA heritage dating as far back as one million years ago, although the reliability of these studies has been questioned by others. It is most commonly believed that competition from *Homo sapiens* probably contributed to Neanderthal extinction, and that the species could have coexisted in Europe for as long as 10,000 years.

H. sapiens (the adjective *sapiens* is Latin for "wise" or "intelligent") proper have lived from about 250,000 years ago to the present. Between 400,000 years ago and the second interglacial period in the Middle Pleistocene, around 250,000 years ago, the trend in skull expansion and the elaboration of stone tool technologies developed, providing evidence for a transition from *H. erectus* to *H. sapiens*.[40] The direct evidence suggests there was a migration of *H. erectus* out of Africa, then a further speciation of *H. sapiens* from *H. erectus* in Africa. A subsequent migration of *H. sapiens* within and out of Africa eventually replaced the earlier dispersed *H. erectus*. This migration and origin theory is usually referred to as the *recent single origin* or, again, the Out of Africa theory. Current evidence does not preclude some multiregional evolution or some mixture of the migrant *H. sapiens* with existing *Homo* populations; this is a hotly debated area of paleo-anthropology.

Current research has established that humans are genetically highly homogenous; that is, the DNA of individuals is more alike

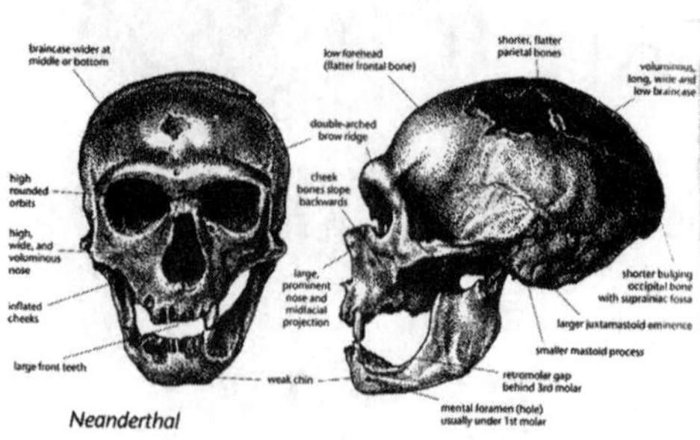

Neanderthal

than is usual for most species, which may have resulted from their relatively recent evolution. Distinctive genetic characteristics have arisen, however, primarily as the result of small groups of people moving into new environmental circumstances. These adapted traits are a very small component of the *Homo sapiens* genome, but include various characteristics such as skin color and nose form, and to internal characteristics such as the ability to breathe more efficiently at high altitudes.

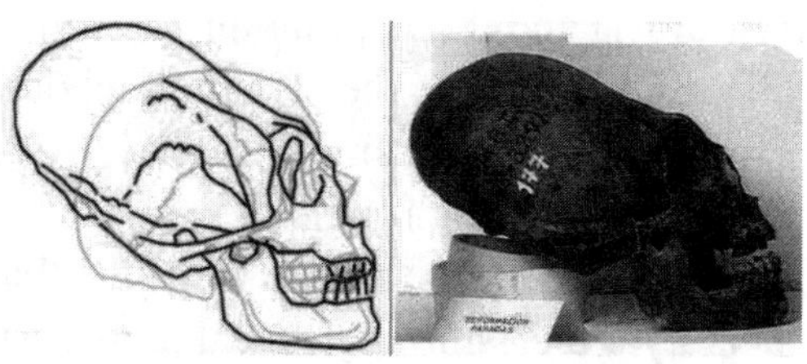

An elongated skull from Ica, Peru.

According to the Out of Africa model, developed by Chris Stringer and Peter Andrews, modern *H. sapiens* evolved in Africa 200,000 years ago. *Homo sapiens* began migrating from Africa between 70,000 and 50,000 years ago, and eventually replaced existing hominid species in Europe and Asia. Out of Africa has gained support from research using mitochondrial DNA (mtDNA).[41] After analyzing genealogy trees constructed using 133 types of mtDNA, researchers concluded that all were descended from a woman from Africa, dubbed "Mitochondrial Eve." Out of Africa is also supported by the fact that mitochondrial genetic diversity is highest among African populations.

Aside from the popular Out of Africa theory, there is also a multiple dispersal model called the *Southern Dispersal* theory, which has gained support in recent years from genetic, linguistic and archaeological evidence. In this theory, there was a coastal dispersal of modern humans from the Horn of Africa around 70,000 years ago. This group helped to populate Southeast Asia and Oceania, explaining the discovery of human sites in these areas much older than those in the Levant (an area of the Middle East often called the Holy Land). A second wave of humans dispersed across the Sinai Peninsula into Asia, resulting in the bulk of the human population in Eurasia. This second group possessed a more sophisticated tool technology and was less dependent on coastal food sources than was the original group. Much of the evidence for the first group's expansion would have been destroyed by the rising sea levels at the end of the Holocene era. The Southern Dispersal model is contradicted by studies indicating that the populations of Eurasia and the populations of Southeast Asia and Oceania are all descended from the same mitochondrial DNA lineages, which evidence supports a single migration out of Africa that gave rise to all non-African populations.[42]

Yet another alternative theory is that of multiregional origins. This hypothesis holds that humans first arose near the beginning of the Pleistocene Epoch about two million years ago, and that subsequent human evolution has been within a single, continuous human species. This species encompasses archaic human forms such as *Homo erectus* and Neanderthals as well as modern forms, and evolved worldwide to the current diverse populations of modern *Homo sapiens sapiens*. The theory contends that humans evolve through a combination of adaptation to the ecology within various regions of the world and gene flow between those regions.

However, aside from all of these other ideas, which I am sure is confusing to the reader, genetic evidence from the late 1980s on the mitochondrial genome indicated that all living humans had as an ancestor a single female living in Africa about 200,000 years ago. This led to a hypothesis that around that time period, a small founder population of humans left Africa and eventually replaced all archaic humans then living outside of Africa without interbreeding, contrary to the multiregional hypothesis. However, as data on the far larger nuclear DNA genome started to become available, evidence mounted that genetic contributions from archaic human populations from around the world, and not just from Africa, also persist in modern humans. However, this work is still in its infancy, and is quite confusing.

So, boiling all the above down gives us a reasonable idea that modern humans, one way or the other, had their genetic roots in Africa, and that all of humanity of the present day came from there. This theory does not take in the possibility of sunken lands in the Atlantic, Pacific or Indian oceans being the origin of mankind, as some mystics have suggested. As all modern humans, dating as back at least 50,000 years according to most accounts are the same as regards anatomy, where could the possibility of people existing with natural elongated skulls fit into this equation?

New DNA testing may show that the many elongated crania

fit neatly into the pattern of all human beings descending from a single female living in Africa 200,000 years ago—or perhaps it will show that there is a completely different race of humans that are descended from some other ancestor. Perhaps that ancestor is literally "from another world." Only time—and further testing—will tell.

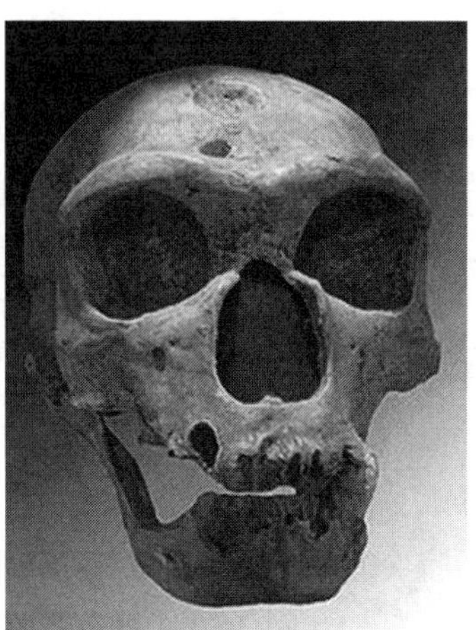

Neanderthal cranium.

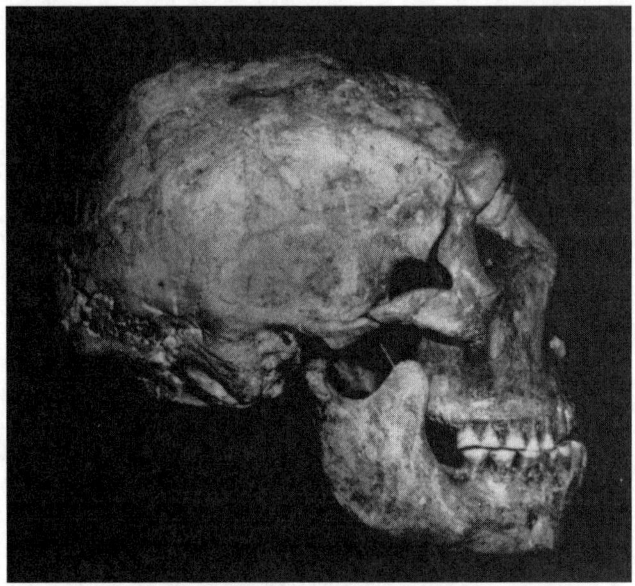

A Neanderthal skull from Shanidar Cave, Iraq.

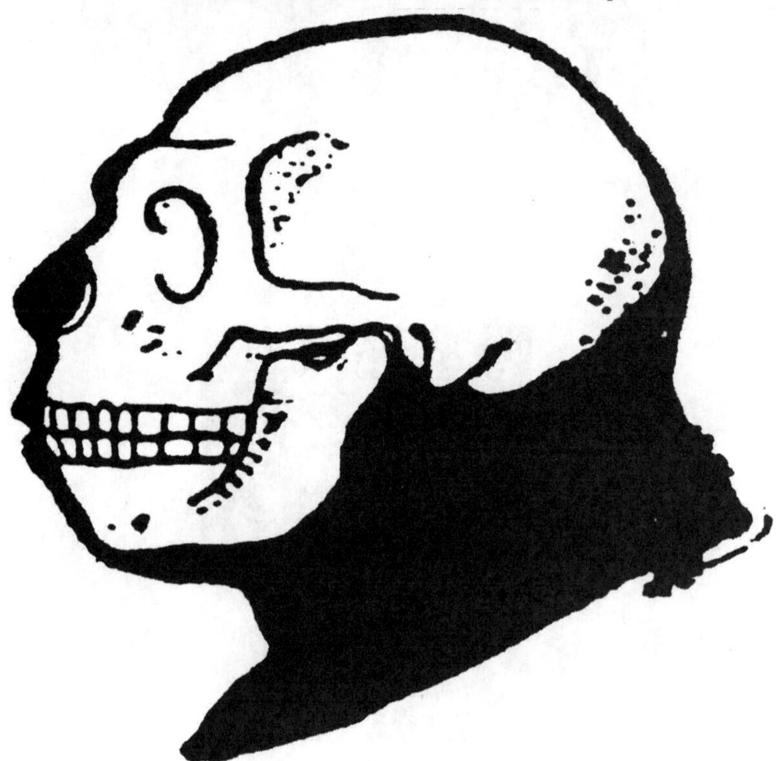

A reconstruction of a Neanderthal cranium.

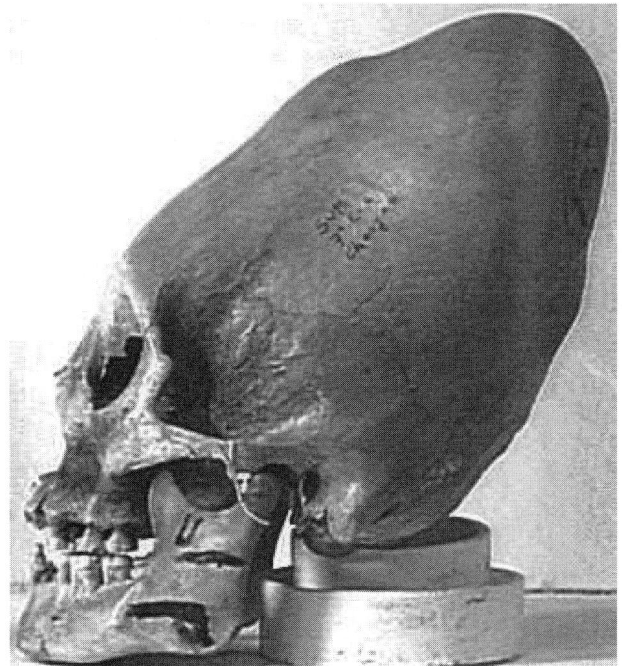

An elongated skull from Ica, Peru.

A Maya skull with an extended cranium.

CHAPTER 3

CRANIAL DEFORMATION AND THE OLMECS

We can trace the progress of man in Mexico
without noting any definite Old World influence during
this period (1000-650 BC) except a strong Negroid substratum
connected with the Magicians (High Priests).
—Frederick Peterson, *Ancient Mexico (1959)*

The Strange World of the Olmecs

The oldest and probably greatest mystery of early Mexico, and North America in general, is the problem of the Olmecs. When one looks at the enigmatic cave drawings, the colossal basalt heads, the trademark "frown," and the elongated heads of most of the Olmecs, an emphatic question leaps to the forebrain: "Who are these weirdos?"

The strange world of the Olmecs is only now being pieced together. In their art, Olmecs are often dressed in leather helmets, have broad faces and thick lips (plus broad noses), and could easily be likened to African kings or rugby players. While mainstream archeologists assure us that Africans never colonized Mexico or Central America, the average man looks at these statues and heads and wonders how they could not be African.

What is fascinating about this enigmatic civilization to us modern viewers is how they represented themselves. In addition to those showing Negroid features, many artifacts depict individuals who have Oriental or European features. It is very interesting to pay close attention to how the figures are presented—how they dressed; the head gear they wore; the shape of their eyes, nose, ears and mouths; the way they held their hands; and the expressions on their faces. It is all wonderful art at its finest. The expressions and symbolism in the objects they hold or are associated with seem to indicate a high level of sophistication, and a shared iconography—What does it all mean? Who are these people? Were they isolated villagers or strangers from a faraway land?

The Discovery of the Olmecs

Until the 1930s it was largely held that the oldest civilization in the Americas was that of the Maya. The great number of Mayan monuments, steles, pottery, statues and other artifacts discovered throughout the Yucatan, Guatemala and the Gulf Coast of Mexico had convinced archeologists that the Maya were the mother civilization of Central America.

But some "Mayan" artifacts were different from the main bulk of the artifacts in subtle ways. One difference was that some carvings of large heads had faces with more African-looking features than many of the other Mayan works. Mayan paintings and sculpture can be quite varied but the African-looking features seemed distinctly un-Mayan. These African-looking heads often had a curious frown and often wore masks or appeared to be a half-jaguar-half-man beast. This recurring motif did not fit in with other Mayan finds.

In 1929, Marshall H. Saville, the Director of the Museum of the American Indian in New York, classified these works as being from an entirely new culture not of Mayan heritage. Somewhat inappropriately, he called this culture Olmec (a name first assigned to it in 1927), which means "rubber people" in Nahuatl, the language of the Mexica ("Aztec") people. Most of the early anomalous artifacts were found in the Tabasco and Veracruz regions of southern Mexico, a swampy region exploited for natural gas, but in ancient times a source point for rubber. Ancient Mesoamericans, spanning from the Olmecs to Aztecs, extracted latex from *Castilla elastica*, a type of rubber tree in the area. The juice of a local vine *Ipomoea alba*, was then mixed with this latex to create rubber as early as 1600 BC (and possibly earlier). "Olmec" was the Aztec name for the people who lived in this area at the much later time of Aztec dominance.

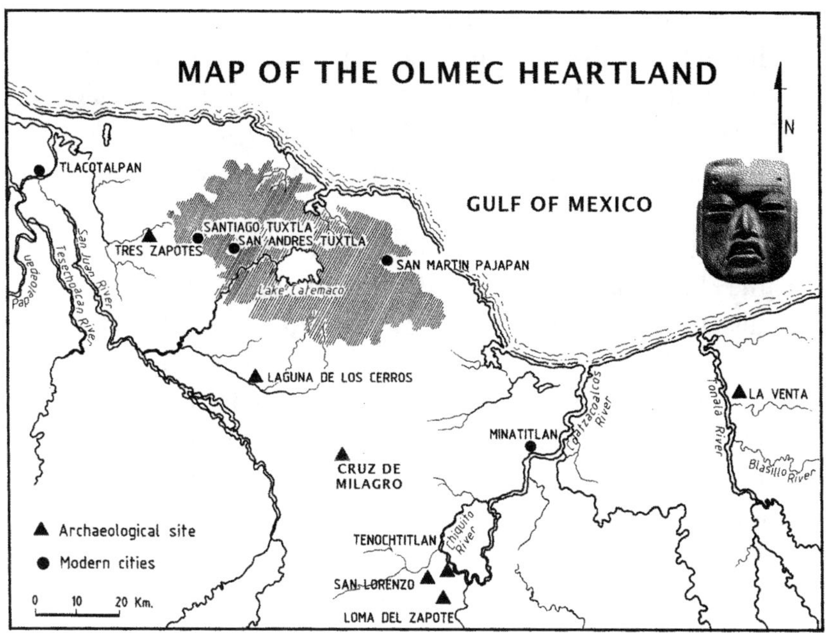

The ancient Olmecs are now credited with creating the ball game that played such a significant role in all Mesoamerican civilizations, and the rubber balls that were used in the game. This game may be even older than the Olmecs, in fact. Ball courts and the Olmec-Mayan ball game were popular as far north as Arizona and Utah and as far south as Costa Rica and Panama.

According to the famous Mexican archeologist Ignacio Bernal, Olmec-type art was first noticed as early as 1869 but, as noted above, the term "Olmec," or "Rubber People," was first used in 1927. Naturally, a number of prominent Mayan archeologists, including Eric Thompson who helped decipher the Mayan calendar, refused to believe that this new culture called the Olmecs could be earlier than the Mayas. Not until a special meeting in Mexico City in 1942 was the matter largely settled that the Olmecs predated the Mayas. The date for the beginning of the Olmec culture was to remain a matter of great debate, however.

Bernal sums up this curious archeological episode in his book *A History of Mexican Archaeology*[1]:

> ...the whole Olmec culture is earlier than the Maya. This was anathema at the time because, as we have already

seen, almost all the endeavors of the Carnegie and other institutions, North American in particular, had been directed towards Maya research, the consensus of opinion then being not only that the Maya culture was the oldest, but that all the other Mesoamerican cultures had stemmed from it.

At the celebrated meeting held under the aegis of the Sociedad Mexicana de Antropología in 1942 to discuss the Olmec problem, archaeologists headed by Caso, Covarrubias and Noguera, along with Stirling, all maintained that the Olmecs belonged within the Archaic horizon. Caso claimed that the Olmec 'is beyond doubt the parent of such other cultures as the Maya, the Teotihuacan and that of El Tajín' (1942:46). Covarrubias held that 'whereas other cultural complexes share "Olmec" traits, this style contains no vestiges or elements taken over from other cultures, unless it be from those known as Archaic' (1943:48). Vaillant was one of the few North Americans to back up these theories and he did so because, in the course of his fine work on the Central Plateau with which we are already familiar, he had come across Archaic figurines displaying undoubted kinship with Olmec types. Eric Thompson, on the other hand, thought that the Olmec was a late culture within what we have now come to call the Post-Classic

And yet the name Olmec, first used by Beyer (1927) to designate this particular art style, has prevailed until today, incorrect though it may be. It is a source of confusion because it is lifted from historical sources which apply the term Olmec to very much later peoples. In 1942 Jiménez Moreno cleared the matter up by showing that the name Olmec properly refers to the inhabitants of the natural rubber-growing areas, but even so we have to distinguish clearly between the relatively recent bearers of the name and the archaeological Olmecs, which is why he proposed that these be called the 'La Venta people' to make confusion less confounded. But the name given at baptism was not to

> be shaken off, and is the one still used today.
>
> At the Mesa Redonda de Tuxtla in 1942 the Olmecs were given a provisional starting date around 300 BC. But somewhat later work at San Lorenzo, carried out with the aid of radiocarbon analysis—the use of which was spreading throughout the area—showed that 1200 BC was a more realistic date. This fitted in perfectly with what was being discovered all over Mesoamerica. It is a part of the general process that has already been discussed. Nineteenth-century scholars had often proposed fabulously early dates for the prehispanic peoples, and it produced in this century a vigorous counter-reaction which in its turn condensed them too drastically. But after 1950 this difficulty was to be overcome by the use of dating techniques that are not essentially archaeological.[1]

The Olmecs had been discovered. However, this discovery created more questions than there were answers. The discovery of the Olmecs seemed to cast into doubt many of the old assumptions concerning the prehistory of the Americas. Suddenly, here was a diverse-looking people who built monumental sculptures with amazing skill, were the actual "inventors" of the number and writing system used by the Maya, the ball game with its rubber

Olmec figurines from Bernal's book, *The Olmec World*.

balls and even knew about the wheel (as evidenced by their wheeled toys).

The greater enigma was upon the archeologists—who were the Olmecs?

Who Were the Olmecs?

Bernal continued to study the Olmecs and came out with the significant study on this early Central American culture in his 1969 book *The Olmec World*.[3] In that book, Bernal discussed the curious finds attributed to the Olmecs all over southern Mexico and Central America, as far south as the site of Guanacaste in Nicaragua. However, he could not figure out the origin of these strange and distinctive people whose art featured bearded men, Negroid heads, and undecipherable hieroglyphs. Even such famous Mayan sites as Uaxactun and El Mirador were thought by Bernal to have been previously occupied by the Olmecs.

The idea that the these strange Negroid heads might be the result of early African exploration seems totally alien to the historians and archeologists who have taken over the archeology of the Americas. Despite depictions of various lords, kings, travelers, magicians and whatnot that look like Africans, Chinese, bearded Europeans, or some other strangers, most professors teaching at our major universities maintain that they are not evidence of ancient pre-Columbian explorers. They admit, though, that people might erroneously get this idea from a "superficial" view of these various statues and carvings.

So, even to mainstream historians, the origin of the Olmecs is a mystery. In the realm of alternative history, many theories

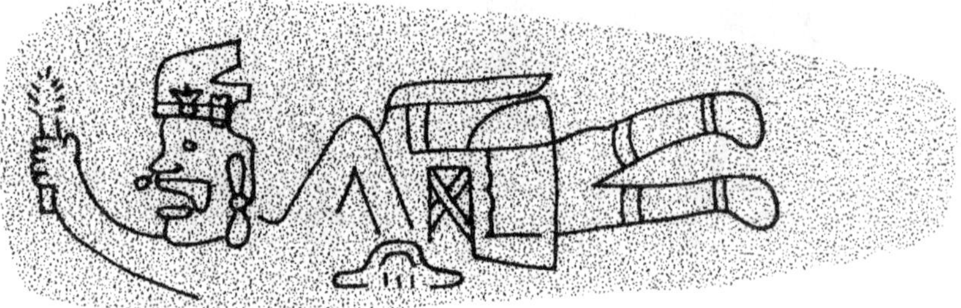

The famous "flying Olmec" carved onto a jade celt, now at the National Museum of Anthropology in Mexico City.

Mathew Stirling clearing debris from the Tres Zapotes Colossal Head 1, 1939.

exist on how Negroids arrived in Central America. One theory is that they are connected with Atlantis; as part of the warrior-class of that civilization they were tough and hard bitten. Or perhaps they were part of an Egyptian colony in Central America or from some unknown African empire. Others have suggested that they came across the Pacific from the lost continent of Mu, or as Shang Chinese mercenaries. Similarly, there is the curious association of "magicians" (or shamanic sorcerers using magic mushrooms and other psychedelics) with many of the Olmec statues—magicians from Africa, China, or even Atlantis?

Are the Olmecs Transoceanic Colonizers?

It is not known what name the ancient Olmec used for themselves; some later Mesoamerican accounts seem to refer to the ancient Olmec as "Tamoanchan." The classic period for the Olmecs is generally considered to be from 1200 BC ending around 400 BC. Early, formative Olmec artifacts are said to go back to 1500 BC, and probably earlier.

Excavations in 1945 at San Lorenzo, Veracruz State.

No one knows where the Olmecs came from, but the two predominant theories are the following:

1. They were Native Americans, derived from the same Siberian stock as most other Native Americans, and just happened to accentuate the Negroid genetic material that was latent in their genes.

2. They were outsiders who immigrated to the Olman area via boat, most likely as sailors or passengers on

> transoceanic voyages that went on for probably hundreds of years.

At the center of the debate about the origin of the Olmecs is the classic struggle between isolationists (who think that ancient man was incapable of transoceanic voyages, and therefore, nearly every ancient culture developed on its own) and diffusionists (who think that ancient man could span the oceans, which explains similarities in widely disparate cultures). There are a few proponents of diffusionism at the traditional academic level. Ivan Van Sertima of Rutgers University in New Jersey actively promotes the diffusionist theory that ancient man crossed both the Atlantic and the Pacific in prolonged transoceanic contact. His books, *African Presence in Early America*4 and *African Presence in Early Asia,*[5] are filled with articles and photos that show without a doubt that Negroes have lived, literally, all over the world, including in the ancient Americas. While Van Sertima does not bring in such unorthodox theories as Atlantis or a lost continent in the Pacific, he is clearly of the belief that Negroes in ancient times developed many advanced civilizations and lived all over the globe.

Unfortunately, most of the writers in the academic field prefer to champion the isolationist theories to the virtual exclusion of the diffusionist.

In the recent scholarly book by Richard A. Diehl, *The Olmecs: America's First Civilization*,[19] Diehl has only one paragraph on the subject, saying:

> The origins of Olmec culture have intrigued scholars and lay people alike since Tres Zapotes Colossal Head I, a gigantic stone human head with vaguely Negroid features, was discovered in Veracruz 140 years ago. Since that time, Olmec culture and art have been attributed to seafaring Africans, Egyptians, Nubians, Phoenicians, Atlanteans, Japanese, Chinese, and other ancient wanderers. As often happens, the truth is infinitely more logical, if less romantic: the Olmecs were Native Americans who created a unique culture in southeastern Mexico's Isthmus of Tehuantepec.

> Archeologists now trace Olmec origins back to pre-Olmec cultures in the region and there is no credible evidence for major intrusions from the outside. Furthermore, not a single *bona fide* artifact of Old World origin has ever appeared in an Olmec archaeological site, or for that matter anywhere else in Mesoamerica.

With this paragraph Diehl summarily dismisses all theories and evidence of transoceanic contact. We don't really know what a *bona fide* artifact would be, since Old World and New World articles were often identical. Also, we are given no further information on the pre-Olmec cultures that the Olmecs are presumably derived from.

But for the Olmecs to actually be Africans—not just look like them—they would almost certainly have come to the Isthmus of Tehuantepec via ship. But since such voyages are dismissed immediately and there will be no further discussion of it, the Olmecs simply have to be local boys who have always pretty much been there. At some time in remote prehistory, their early genetic group walked into this Olmec heartland area.

According to Diehl, the Olmecs would have been an isolated group within their region as well, with little contact with other tribes in the Isthmus of Tehuantepec. Says Diehl:

> We do not know what these people called themselves, or if they even had a term that encompassed all the inhabitants of Olman. There is no evidence that they formed a single unified ethnic group, and almost certainly no Olmec considered people living more than a few hours' walk away as members of his or her own group. Nevertheless, the numerous independent local cultures were so similar to one another that modern scientists consider them a single generic culture.

The Olmec settlements, according to Diehl, rose up independently in their corner of Mesoamerica without the influence of any other culture. They all suddenly began making monumental

Ignacio Bernal at Altar Number 4, La Venta, circa 1967.

statues out of basalt (one of the hardest and most difficult stones to carve), and made large structures with sophisticated drainage systems. But they weren't really in contact with their early neighbors. The spread of Olmec-like artifacts was achieved only later when Olmec "styles" were used by other more widespread cultures.

Diehl was actually proved wrong on this account when it was announced in January 2007 that an Olmec-influenced city had been found near Cuernavaca, hundreds of miles from the Olmecs' Gulf Coast territory, at Zazacatla. Announcements at that time stated that a "2,500-year-old city influenced by the Olmecs, often referred to as the mother culture of Mesoamerica, has been discovered hundreds of miles away from the Olmecs' Gulf coast territory, archaeologists said." (*National Geographic News*, January 26, 2007) This discovery is not really surprising since the Olmec city of Chalcatzingo near Mexico City was excavated and written about in the 1970s.

So, the preponderance of the evidence shows that the Olmecs were very aware of the villages near to them, and aware of cities and peoples quite far from them. Were they aware of transoceanic civilizations as well?

Diehl's book, already outdated though published in 2004, is an interesting but fairly dry read. Not only does he refuse to discuss Negroid features and transoceanic contact in his book, but, except for one brief mention (below) there is no discussion of cranial deformation at all, which is one of the more curious habits of the Olmecs and the Maya, and is found among many other cultures worldwide.

The Olmecs had many unusual similarities with the Maya and other transoceanic cultures, such as: the reverence for jade and exotic feathers; the use of hallucinogenic mushrooms and other psychedelic drugs; the use of hieroglyphs on stone stelae as markers; and the practice of cranial deformation.

Even though cranial deformation is obviously a major feature of the Olmecs, Diehl only mentions it once, in passing, in his book. Says Diehl of artifacts and skeleton found at an Olmec burial at Tlatilco:

> One high-status woman was laid to rest with 15 pots, 20 clay figurines, 2 pieces of red-painted bright green jadite that may have formed part of a bracelet, a crystalline hematite plaque, a bone fragment with traces of alfresco paint, and miscellaneous stones. Another burial held the remains of a male whose skull had been deliberately modified in infancy and whose teeth were trimmed into geometric patterns as an adult. He may have been a shaman since all the objects placed with him were likely part of a shaman's power bundle. They included small *metates* for grinding hallucinogenic mushrooms, clay effigies of mushrooms, quartz, graphite, pitch, and other exotic materials that could have been used in curing rituals. A magnificent ceramic bottle placed in his grave depicted a contortionist or acrobat who rests on his stomach with his hands supporting his chin while his legs bend completely around so that his feet touch the top of his head. Could this masterpiece be an effigy of the actual occupant of the

grave?

Indeed, Diehl almost gets excited about the Olmecs. Could they actually be psychedelic jaguar shamans who like to make monumental heads to keep themselves busy?

The Olmecs, by any standard, are a fantastic, amazing, confusing, psychedelic, and in some cases, just plain weird people. We do not know where they came from. We don't know why they were there. We don't know what their "mission" was. In short we know very little about them. All we really know is that they are old, and they are strange.

While it is easy to see them as Proto-Mayans and Citizens of Olman (however large that country might have been), we should also consider them as the fantastic Proto-Mesoamericans they may have been: psychedelic aliens who used lasers to cut colossal basalt heads; Atlantean refugees who made a last stand in Tabasco; or Shang Chinese mercenaries taken from East Africa or Melanesia and specially trained to administer the Pacific (and later

Mathew Stirling (center) and colleagues at one of the colossal heads at San Lorenzo, 1946. Its eye had appeared in an eroded trail, leading to the discovery of this early Olmec city.

Atlantic) ports of the Isthmus of Tehuantepec; or perhaps a people originally from the Atlantic side all along, having come from Africa, possibly as a military force from Egypt or West Africa, circa 1500 BC. There are many possibilities.

So, with an open mind, let us look into the mysteries of the Olmecs, their fantastic art, their sophisticated technology, their unusual number system, writing system and other customs. What we find may surprise us. We are likely to find that the Olmecs were geniuses, but many other cultures had surprisingly similar ideas.

Olman: The Land of the Olmecs

The Olmecs are said to have occupied "The Land of Olman." This was a designation that the Aztecs used to describe the jungle areas of the nearby coast. The traditional definition of the Olmecs is that they were an ancient Pre-Columbian people living in the tropical lowlands of south-central Mexico, roughly in what are the modern-day states of Veracruz and Tabasco on the Isthmus of Tehuantepec. Their immediate cultural influence went much further though, as their artwork has been found as far away as El Salvador and Costa Rica.

The Olmec heartland is thought to be an area on the Gulf of Mexico coastal plain of southern Veracruz and Tabasco. This is because the area boasts the greatest concentration of Olmec monuments as well as the greatest number of Olmec sites. This area is thought to be the most northerly area of the Mayan realms, with such sites as Comacalco as the northernmost Mayan city along the Gulf Coast of the Ithmus of Tehuantepec.

The Olmec heartland is characterized by swampy lowlands punctuated by low hill ridges and volcanoes. The Tuxtla Mountains rise sharply in the north, along the Bay of Campeche. Here the Olmecs constructed permanent city-temple complexes at several locations: San Lorenzo Tenochtitlán (usually just referred to as just San Lorenzo), Laguna de los Cerros, Tres Zapotes, La Mojarra and La Venta.

The heartland of the Olmecs is also the narrowest land area in Mexico, an area extremely important if an ocean-to-ocean trade route were to be established. This narrow area of southern Mexico is

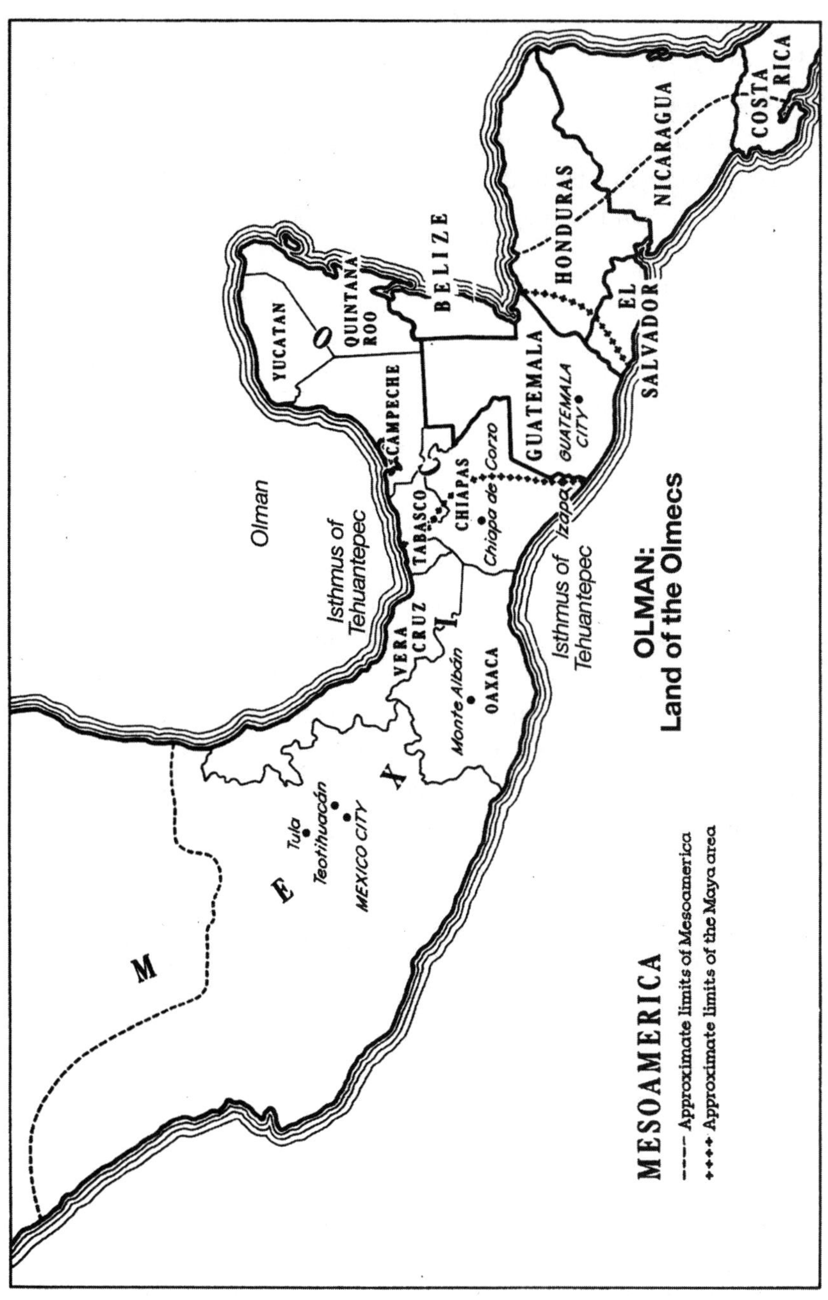
MESOAMERICA
Approximate limits of Mesoamerica
Approximate limits of the Maya area
OLMAN:
Land of the Olmecs
Olman
M
E
X
Tula
Teotihuacán
MEXICO CITY
Monte Albán
OAXACA
VERA CRUZ
TABASCO
CAMPECHE
YUCATAN
QUINTANA ROO
BELIZE
CHIAPAS
Chiapa de Corzo
Izapa
GUATEMALA
GUATEMALA CITY
EL SALVADOR
HONDURAS
NICARAGUA
COSTA RICA
Isthmus of Tehuantepec
Isthmus of Tehuantepec

known as the Isthmus of Tehuantepec and it represents the shortest distance between the Gulf of Mexico and the Pacific Ocean. The name comes from the town of Santo Domingo Tehuantepec in the state of Oaxaca, which in turn comes from the Nahuatl *tecuani-tepec* ("jaguar hill").[6]

The Olmecs were able to link both the Atlantic and Pacific ports through the Isthmus of Tehuantepec, an area that was the original heartland of the Olmecs.

Ignacio Bernal has this to say in his book, *The Olmec World*:

> The Isthmus of Tehuantepec unites the Olmec area with the Chiapas depression and the Pacific watershed. It is a region which attracted the Olmecs, for no mountains cut across it and the climate was tropical. In the central depression, and generally speaking, in the entire state of Chiapas, Olmec remains or others related to it appear constantly, though—as in other regions of Central America—they do not constitute the basis or majority of archaeological finds. We are dealing with a culture related to the Olmec, though with its own peculiar features.
>
> Black ceramics with white rims or spots appear quite frequently. At other sites such as San Agustin and on the Pacific coast of Chiapas the same ware has been found in scientific explorations. At Santa Cruz it is clearly associated with other types belonging to the Olmec complex. At Mirador abundant Olmec figurines have been unearthed.[3]

What Bernal is trying to establish here is that Olmecs were not just on the Atlantic coast but also on the Pacific coast of Chiapas. Olmecs are now known to have been on the Pacific coasts of Guatemala and El Salvador, too.

Olmecs in Lower Central America

The Olmecs are also known to have occupied, or at least influenced, large areas of lower Central America, from Guatemala and El Salvador to Nicaragua, Costa Rica and probably beyond.

One of the most famous statues in the Nacional Museum in San Jose, Costa Rica is an Olmec hunchback figure with an elongated cranium and oriental-type Olmec eyes. Costa Rica is also the site of the enigmatic, perfectly formed granite balls that defy explanation. Were they made by the Olmecs in a similar manner as the colossal heads?

Given that sites like Tonalá and Izapa were early Olmec sites that were later occupied by the Maya, other sites such as Monte Alban further north towards the Valley of Mexico can be assumed to have been first inhabited by the Olmec and then by later cultures.

Once the Olmecs had been established as the oldest culture in Mesoamerica in the 1940s, by default they became the founders of many of the ancient cities. Essentially, if it could be proven that Olmec iconography was being used at an archeological site, then it must have been the Olmecs who founded that city, since the Olmecs are the oldest major culture.

Since the oldest Maya sites such as Uaxactun in the Peten jungles north of Tikal are thought to have been first built by the Olmecs, it is possible that other older Mayan sites were also founded by the Olmecs. This list of Mayan sites founded by the

Excavations in the 1940s at La Venta.

Olmecs could include Copan, El Mirador, Piedras Negras and many others.

La Venta and the Olmec Heartland

Perhaps the Olmec capital was at La Venta, one of the greatest and most famous of the Olmec sites, though now mostly destroyed. The site is typically dated to have been active between 1200 BC and 400 BC, which places the major development of the city in the so-called Middle Formative Period. Located on an island in a coastal swamp overlooking the then-active Río Palma river, the city of La Venta would have controlled a region between the Mezcalapa and Coatzacoalcos rivers. Today, the Rio Palma is an inactive river, the area having reverted into a mass of swamps. One wonders if the Olmecs, as part of a gigantic earthworks project, had created a river through the swamp in which they created their "capital," if that is what it was.

The site of La Venta itself is about 18 miles inland with the island consisting of slightly more than two square miles of dry land. The main part of the site is a complex of clay constructions stretched out for 12 miles in a north-south direction, although the site is 8° West of true North. The entire southern end of the site is now covered by a petroleum refinery. Most of this section of the city has been largely demolished, making excavations difficult or impossible.

Many of the site's fabulous monuments are now on display in the archaeological museum and park in the city of Villahermosa, Tabasco, the oil capital of Mexico. La Venta and nearby San Lorenzo have been the source of many of the colossal heads that the Olmecs are so famous for. The important basalt quarries for the colossal stone heads and prismatic basalt logs are found in the Tuxtlas Mountains nearby.

Marking the southern end of La Venta's ceremonial precinct is an enormous pyramidal mound. Standing at the base of this pyramidal mound was Stela 25/26. This stela depicts a bundled zoomorphic creature with foliage at the top that is thought to represent a World Tree or axis mundi.

The northern end of Complex A is mainly an enclosed

courtyard, with a massive underground serpentine deposit. This serpentine deposit is thought to represent the primordial waters of creation. Buried beneath the enclosed courtyard was Offering 4, a now famous funeral offering that is an arrangement of six jade celts (adzes) and 15 jade figures of Olmecs with elongated craniums and oriental-looking eyes. A single figure that faces the others is carved from granite. The entire group of figures stand together amongst the upright jade celts that are apparently representing in miniature the tall granite stelae that were commonly used by the Olmecs and Maya (as well as Egyptians, Hindus, and other cultures).

Although the significance of this miniature funeral arrangement is not known, it eloquently demonstrates the conceptual relationship between the forms of the celt-adze and granite stelae, by making the jade celts into miniature stelae with the jade figurines—each with an elongated skull—standing around the celt-stelae, as if at an important meeting. This exquisite arrangement can now be seen at the National Anthropology Museum in Mexico City and is one of the most famous displays in the Olmec section.

Also found at La Venta is the famous Altar 4, which probably functioned as a throne. This massive piece of carved basalt, weighing tons, depicts a ruler wearing a bird headdress and seated within a niche. He holds onto a rope that stretches around to the sides of the altar. On the side of the altar that has not been defaced is a seated individual whose hands are bound by the rope, seemingly as a captive. Another suggestion is, perhaps, that it represents ancestral lineage. Above the seated ruler on the front of the altar is the enormous open maw of a feline creature. This gaping jaguar mouth appears to be metaphorically related to the open portal from which the ruler is emerging.[6]

While it is thought that La Venta was the "capital" or most important city of the Olmecs, this may not actually be the case. We know so little about the Olmecs that it is impossible to say for sure how important La Venta was, or whether there were not more important cities and ceremonial sites for the Olmecs.

For instance, some Olmec sites could be underwater in the Gulf of Mexico, or still buried in the swamps of Tabasco and Veracruz. Or, major Olmec sites might have been located in the interior of Mexico like Chalcatzingo or the recently discovered Zazacatla, nearby. These sites are are quite a distance from the so-called Olmec Heartland, and suggest that the Olmec lands—Olman—were quite extensive. As noted above, sites such as Monte Alban and Teotihuacan are thought by some archeologists to be associated with the Olmecs and it may be the case that these cities were originally important Olmec centers.

The more we find out about the Olmecs, the deeper the mystery surrounding them becomes. We find that the Olmecs seem to include nearly every racial type in the world. How is this possible? The Olmecs are credited with everything from inventing the wheel, the ballgame and hieroglyphic writing, and it is now known they controlled most of southern Mexico from shore to shore. From a diffusionist point of view, the Land of Olman may well have been the "center of the world" as the Ithmus of Tehuantepec would indeed have been the center of the world if there was a strong transoceanic trade across both the Atlantic and Pacific Oceans. If such a trade and movement of ships had existed, the Olmecs may well have been a cosmopolitical center where worldwide cultures intermingled.

The Olmecs and Cranial Deformation

Among the many strange things about the Olmecs, cranial deformation is one of the most bizarre. It is a practice that was shared with the peoples of Peru, some Pacific islands, tribes in Africa, the Kurds of Iraq and the Egyptians—as well as many other cultures. It is an unusual custom that spans both the Atlantic and Pacific Oceans.

Anthropologists have long known that cranial deformation can be accomplished while humans are still infants. Infants would have their heads bound with pieces of wood and rope, or some sort of constrictive cloth, that forced their heads to grow in an elongated and unnatural way.

The skull is bound while the plates in a baby's skull have not yet fused together, and the materials are adjusted and tightened as time goes on. After two or three years of this skull binding, the child's head is permanently growing in an elongated fashion, and

Olmec jade birdman with elongated skull from Costa Rica.

once the child becomes several years old the skull has formed in an elongated fashion that no longer requires any binding.

It is known that the Olmecs and the Mayas of Central America had dolichocephalous crania in many cases. Though skeletal material is often difficult to find in tropical areas because of the fast rate of decomposition, jade figurines attributed to the Olmecs have been found that show persons with elongated heads, so we know that this curious "fashion" was widespread, found from the Andes to the jungles of Panama, Costa Rica and Nicaragua, to the coasts and mountains of Mexico. It even reached the Pacific Northwest of the United States.

Remember, University of Alabama archeologist Richard Diehl says that the Olmecs were an isolated group within their region with little contact even with other tribes in the Isthmus of Tehuantepec, and Diehl's book mentions a cranial deformation in an Olmec burial at Tlatilco. So, apparently, according to Diehl, the "numerous independent local cultures" only a few hours' walk from each other must have each decided, quite without the knowledge that that anyone else was doing it, to bind up their babies' heads until they were grossly misshapen.

Perhaps Diehl's (and others) the lack of discussion of cranial deformation is an actual avoidance of the topic. While little is really known about cranial deformation, except how it was done, it is a curious part of Olmec and Mayan culture, a custom that they are well known for, as are the ancient Peruvians along the desert coast, especially around Paracas, Ica and Nazca, south of Lima. It was also practiced at Tiahuanaco high in the Andes. One would think this strange practice would arouse curiosity and speculation, but since the Olmecs are said to be an isolated group, mentioning cranial deformation might just invite discussion of its widespread use, and prove counter-productive in a book promoting the isolationist point of view.

Cranial deformation, because it is so unusual and found worldwide, would seem to be clear evidence of transoceanic contact. Not only was cranial deformation popular throughout the Americas but it was done all over the world. The popularity of this rather bizarre cultural practice would seem to arise from cultural diffusion, not cultures isolated from each other.

Why Did the Olmecs Deform their Skulls?

Theories, as with all cultures, include such ideas that they thought that bigger heads meant smarter people; that it was a form of imitation of "visitors" who had elongated crania; that it was part of some oddball religious practice; that it was to distinguish a priestly and royal elite; or a combination of some or all of these theories.

The "technology" of deforming crania was relatively simple and any family that had a newborn infant and could get the "tools": a couple of pieces of wood and some rope. All it really took was some patience and time. But once again we have to ask ourselves—why would this be the style?

Central to any unified theory on cranial deformation is the fascinating fact that this was a style that spread to remote corners

Olmec ceramic figure with extended cranium, from Las Bocas, Mexcio, now in the Museum of the American Indian, New York.

of the earth. The Land of Olman on the Isthmus of Tehauntepec would have actually been a central area for such diffusion. Was there a time thousands of years ago when priests around the world all had elongated crania? By sheer "coincidence" this would appear so, as we will see. Cranial deformation was apparently a widespread fad thousands of years ago, lasting up until modern times, and the Olmecs were known to have taken part in it.

The Strange V-Form Cleft

The most bizarre topic when it comes to the Olmecs is that of the V-form cleft seen in some of the statues.

Ignacio Bernal has this to say while discussing jade figurines found at La Venta in his 1969 book *The Olmec World*:[3]

> The deformations which the Olmecs practiced on their own bodies (whether actual or only symbolical) are more clearly visible in the figurines than in the stone monuments.
>
> The head was deformed by binding a small board in oblique position to the forehead of the newborn child until pressure gave the still-plastic cranium the desired form. Circular deformation was carried out also, and it has been suggested that the helmets on the colossal heads or on other figurines represent molds used to produce deformation. One of the most characteristic traits is a cleft in the form of a V in the upper part of the head, which at times plainly becomes a hole. As this appears also in the figures of jaguars it may indicate the deep indention in the top of the skull of the animal. Formerly it was thought that this cleft on human figures represented the type of cranial deformation called *bilobé* which was believed to be practiced by some groups. But it has since been demonstrated that this deformation is anatomically impossible and could not have been practiced by the Olmecs or any other group.
>
> The cleft, possibly representing some real feature, was transformed into a stylistic element which lost its significance even for its makers, although they continued to

> represent it over a long period in celts as well as in figurines. Teotihuacan II figurines at times show this trait. On the other hand, it does not appear at Monte Alban, in spite of the fact that many Olmec vestiges, particularly the jaguar, can be seen there. Thus I believe that if the cleft actually came into existence in representing the supraorbital cavities of the jaguar, the association was lost by the time the jaguar was represented at Monte Alban, where—if this animal is identical to the Olmec beast—it does not have the V-form cleft. The V cut sometimes is shown as a furrow crossing the head, as in a small sculpture found at El Tejar, but this site is outside the Metropolitan Olmec entity. On other figurines the heads are shaved or have only a roll of hair down the center.[3]

Bernal's interesting comment on the V-form cleft as an imitation of the jaguar skull may provide an alternative answer to the worldwide practice of cranial deformation, though it appears that Bernal did not know that cranial deformation was so widespread.

Bernal mentions in a footnote that fellow archeolgist Michael Coe wrote to him in a personal letter: "This cleft has long puzzled me. I would now agree with you that it is a natural feature of the living jaguar, for on a recent visit to the Regent's Park and Washington zoos I have noted that on full-grown male jaguars there is a pronounced furrow on the top of the head, where the rather loose scalp has been almost folded over."

The idea that cranial deformation and the V-form cleft were disfigurements meant to make someone appear more jaguar-like does make some sense, at least for the Olmecs. It is known that they had a were-jaguar cult, wore jaguar masks, and otherwise saw the animal as symbol of power.

Also, the idea that some Olmecs actually had a deep V-form cleft in their heads is rather fantastic—and Bernal quickly dismisses the notion. But it seems that he is on to something. The

V-form cleft would seem to be connected to the head binding and cranial deformation that the Olmecs found so popular.[6]

Did the frowning parents of these babies actually make a deep V depression in the child's head and then bind it? If they did, would it have compressed an area of the brain that made them violent and war-like? Bernal actually mentions the subject, but then backs off. Were subcultures of the Olmecs trying to make themselves actually look like jaguars, with filed teeth, special facial scarring and tattoos, nose and ear piercings to add "jaguar appendages" and other embellishments to the body?

If you think this idea is too fantastic, I refer you to modern-day cat people such as Seattle resident Dennis Avner, known as the Catman, who has had a number of surgeries to make himself look like a cat. The *Seattle Times* reported on him on August 16, 2005 saying that Avner, "who goes by his Native American name, Stalking Cat, is known around the world as the Catman. Over the past 25 years, Stalking Cat, 47, has received so many surgical and cosmetic procedures he's lost count. And he says all of them — from full-face tattoos to fanged dentures to steel implants for detachable 'whiskers'—have been done to achieve oneness with what he calls his totem, the tiger."

The acidic soil around the Gulf Coast Olmec areas has meant that few Olmec burials would preserve the skulls and other bones. So, most Olmec portraits of cranial deformation have come to us from statues and figurines. These indeed show that many of the Olmecs exhibited some sort of cranial deformation, and some of these showed a V-form cleft at the top of the head. Could these figurines actually depict real persons? Bernal, and most archeologists, think not.

Many strange and extremely elongated crania can be seen in museums along the desert coasts of Peru and Chile where they have been well preserved. None of these show a V-form cleft on the cranium. The V-form cleft figurines may possibly be representing the other version of cranial deformation where a child's head is bound in such a fashion as to flatten the head and widen it, which makes a small V at the top.

When discussing cranial deformation and were-jaguars it

seems sensible to ask if the hallucinogenic mushroom trade had anything to do with these customs. Hallucinogenic mushroom rites are well known in Olmec culture, so the question arises, did the Olmec priests, presumably with cranial deformations, ingest magic mushrooms and then "become" were-jaguars in their altered state? It would seem that some sort of similar ritual scenario was going on over much of the Land of Olman.

Modern archeologists are faced with the puzzle that is the Olmecs, and their strange looks and behavior. Olmec figurines give us a wide variety of head and body types that are curious by any historian's standard. It would seem that the Olmecs, with their V-form cleft and other traits, were associated with ancient China and Africa, among other areas. We will explore this more in upcoming chapters.

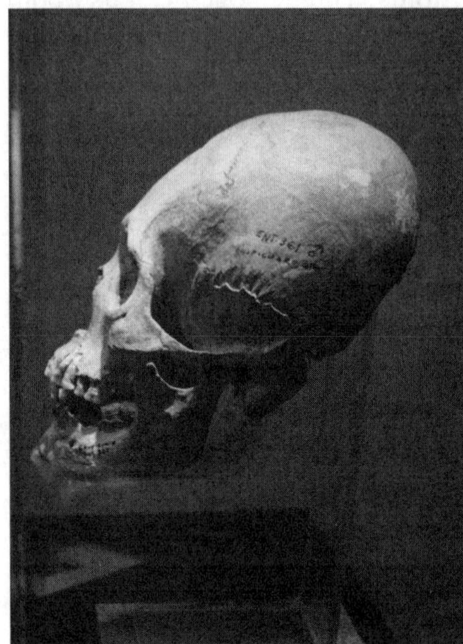

Left and below: Olmec elongated skulls on display at the Anthropology Museum in Xalapa.

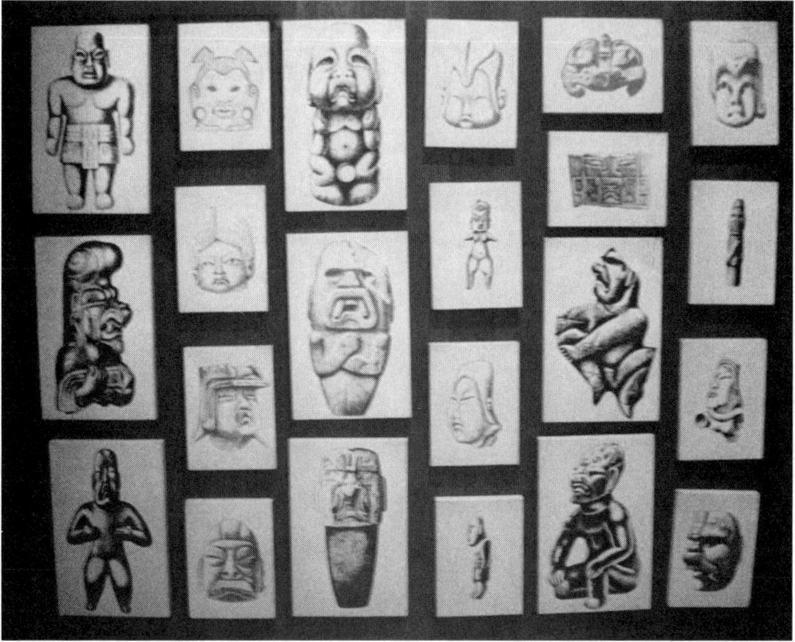

Top: Olmec figurines in clay from Comacalco, on the Mexican Gulf Coast showing elongated heads. Below: A chart from Comacalco illustrating many of the curious figurines discovered at Comacalco, presumed to be Olmec and Mayan.

Aerial view of the La Venta site, circa 1966.

Basalt column tomb discovered in the1940s at La Venta.

Excavations in the 1940s at La Venta, basalt tomb in background.

Jade celt from La Venta with the curious V-form cleft.

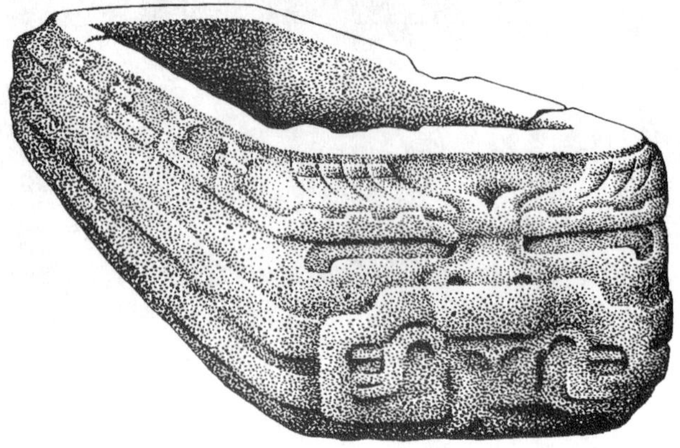

Sarcophagus with were-jaguar face, La Venta.

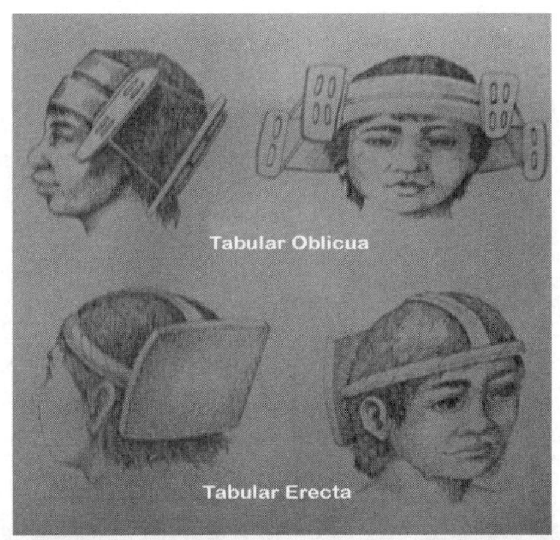

Olmec jadeite votive axe now in the British Museum. It has the V-cleft forehead and typical "flame" eyebrows and Olmec frown, sometimes described as the feline frown of the were-jaguar.

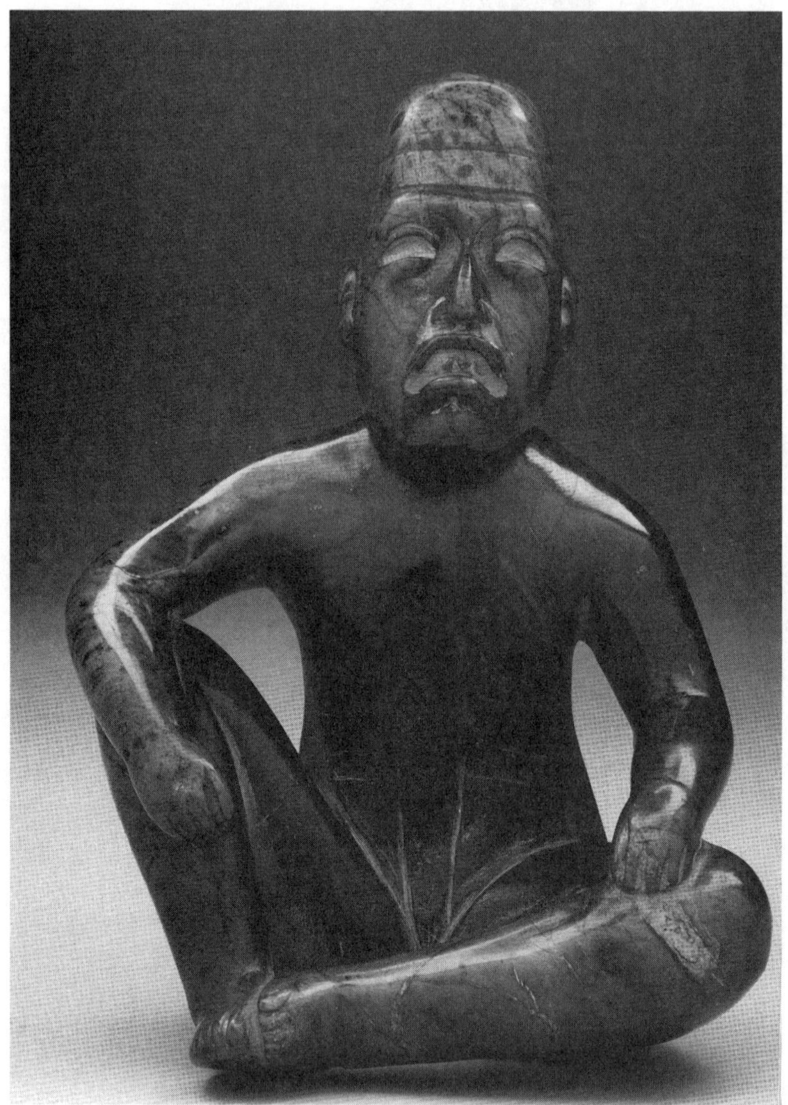

Olmec jade statuette from La Venta with signs of cranial deformation.

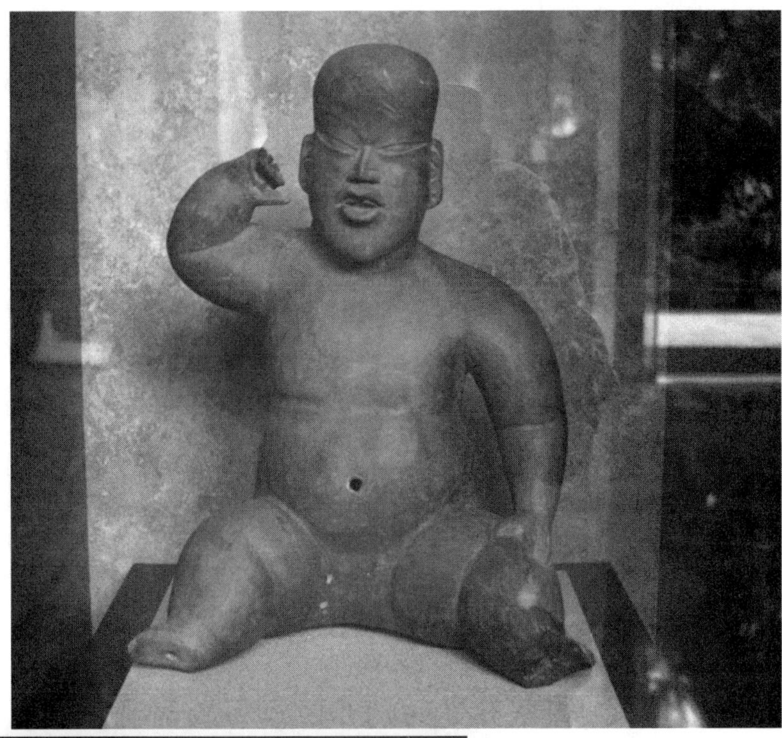

Top: Olmec eunuch baby clay figurine from Tlapacoya, Mexico. *Left*: Olmec eunuch baby ceramic. Both statues show an elongated cranium and are on display at the National Anthropology Museum in Mexico City.

Curious clay Olmec figurines with Asian faces at the Xalapa Museum.

Olmec jade maskette with the curious V-form cleft.

Jade figurine from La Venta known as Offering 3. It seems to show cranial deformation, piercing, facial and body scarring plus tattooing.

Olmec jade figurines with jade celts discovered at La Venta. All of the figures have elongated heads and oriental-type eyes. These priceless jade arifacts are now at the Anthropology Museum in Mexico City.

San Lorenzo Olmec statue apparently with a fake beard and Egyptian-looking headdress.

How the Olmec jade figurines and celts were originally discovered at La Venta. We can pretty much be certain that this is how many of the Olmecs actually looked.

Left: Two Mayan skulls on display at the Archeology Museum at Merida, Mexico. The bottom skull has been flattened into a "Pumpkin-head."

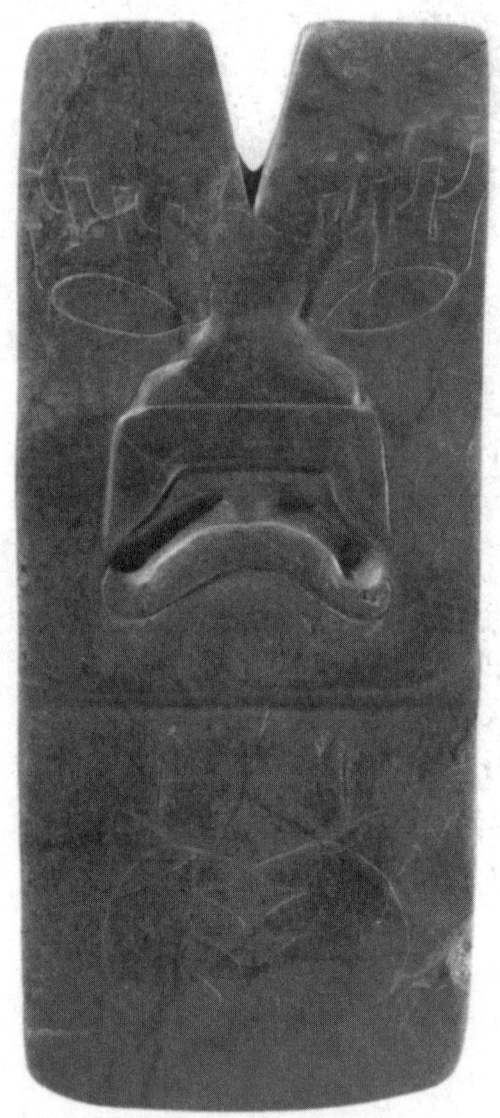

Olmec votive jade axe with V-form cleft from La Venta.

Olmec ceremonial axe carved from green serpentine. It includes the trademark frown and V-form cleft.

Above: Chart showing variations of the V-form cleft and were-jaguar face. *Right*: Monument from San Martin Pajapan with curious double mask and V-form cleft.

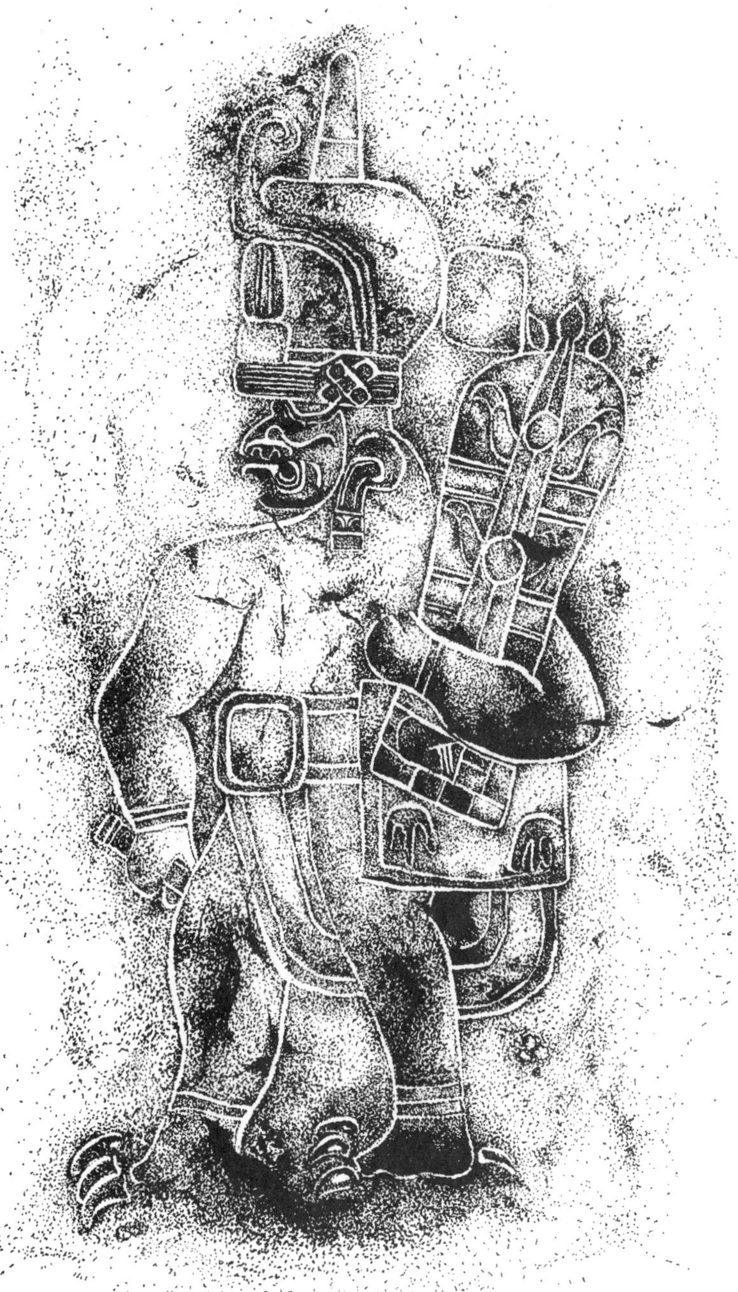

Olmec Were-Jaguar with bird beak from Xoc, Chiapas.

Top: La Venta Altar 5 showing an Olmec emerging from a cave with a baby (a eunuch?). *Bottom*: South side of the monument showing people with elongated heads.

Closeup of south side of La Venta Altar 5. It clearly shows an infant with an elongated head, and we presume that the adults are also coneheads.

A basalt Olmec statue from La Venta showing a person with a "Mohawk" hairstyle —or possible skull deformation. This figure wears the typical Olmec frown, plus he is in the "Quizuo" position, on his knees in a supplicant manner to a superior.

Mayan coneheads depicted in a mural at Bonampak.

Mayan coneheads depicted on a polycrhome vase from Yaxchilan.

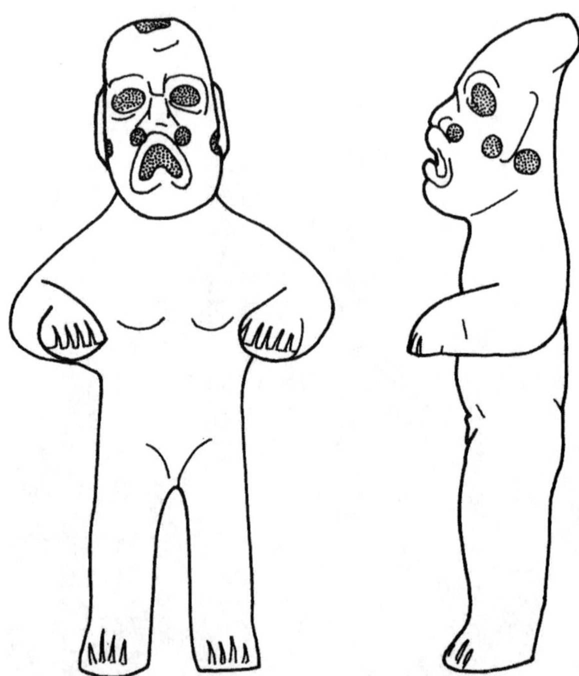

An Olmec figurine from Chalcatzingo, Mexico, with an extended cranium.

Two Olmec figures with elongated heads are attacked by jaguars in this painted scene at the Olmec site of Chalcatzingo in central Mexico.

Stele D from the Olmec site of Tres Zapotes, Mexico. Note the elaborate headresses and the man with a beard. Coneheads?

Guatemalan archeologist Raphael Girard at Olmec excavations in El Salvador.

San Lorenzo, Monument 4, now in the Xalapa Museum.

Tres Zapotes colossal head, first discovered in 1860.

Pockmarked head from San Lorenzo, now at the Xalapa Anthropology Museum. It is a mystery how and why these man-made pockmarks were bored into the hard basalt statue sometime after it was originally carved. It seems like some sort of power tool was used to scoop out the very hard rock and create the even circular marks. This colossal head weighs about 19 tons.

La Venta Monument 19, now in the archeological park in Villahermosa. An Olmec, wearing a helmet and holding a satchel in front of him, sits within the coils of a rattlesnake with a feathered crest, making this the oldest depiction yet found of the "feathered serpent." Does the mysterious satchel hold sacred jade oracle sticks, dried magic mushrooms or something else of great ceremonial importance?

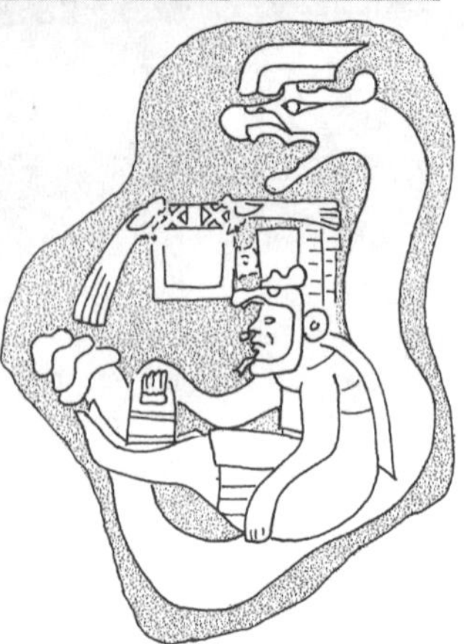

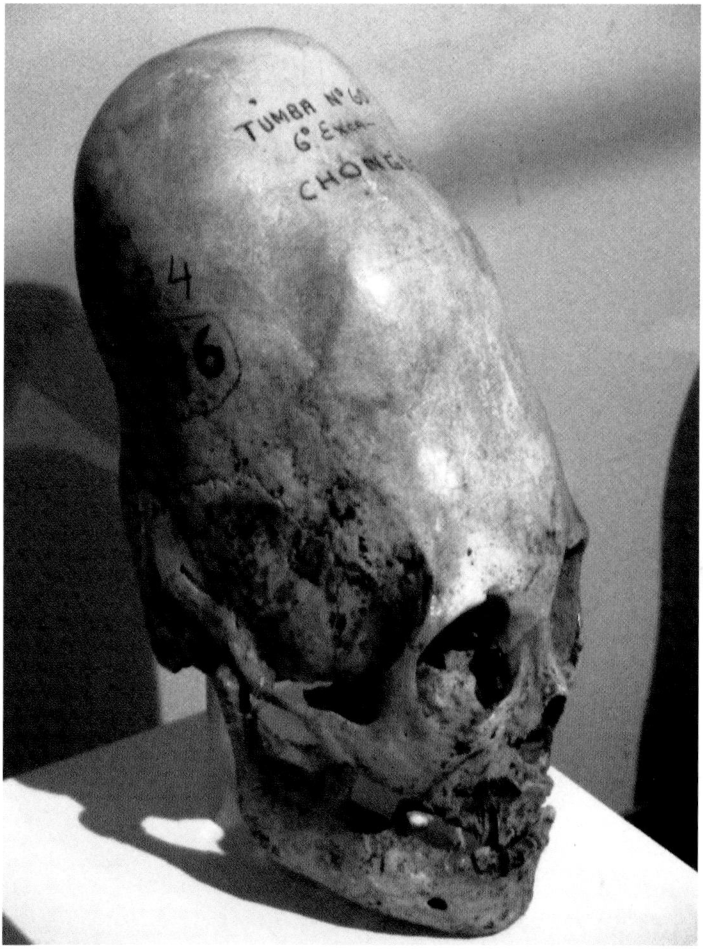

Above: The extended cranium on the famous Chongos skull that was discovered near Paracas, Peru. *Below*: A colorful panel from the Mayan city of Bonampak showing the coneheads of the time.

Above left: Ceramic Olmec baby statuette now at the Xalapa Museum. *Above right*: Olmec skull at the National Museum in Mexico City. Below right: Mayan skulls at the Archeological Musem in Merida, Yucatan.

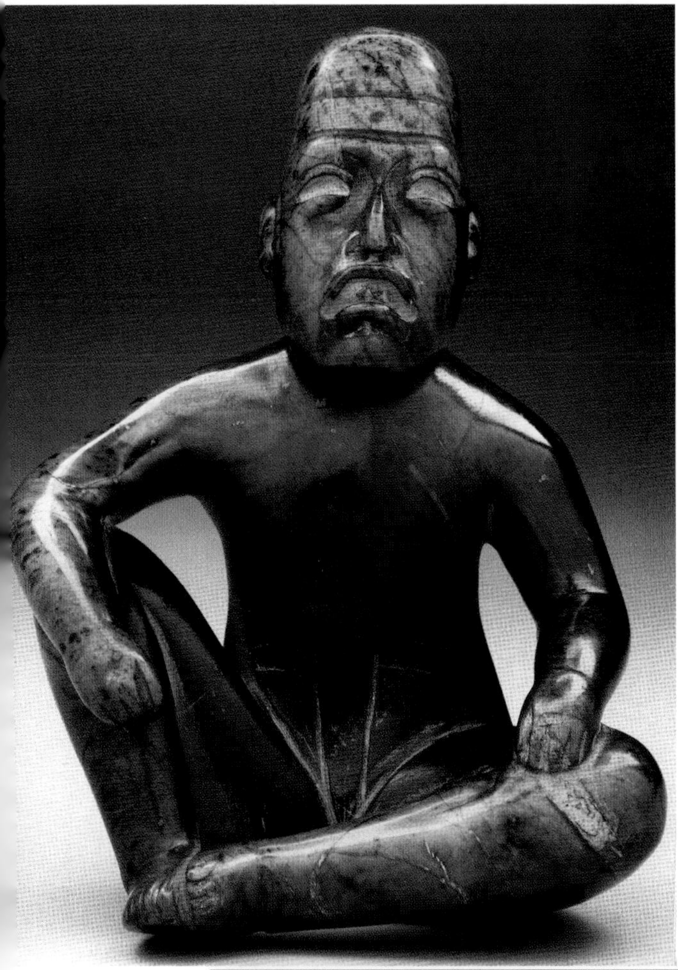

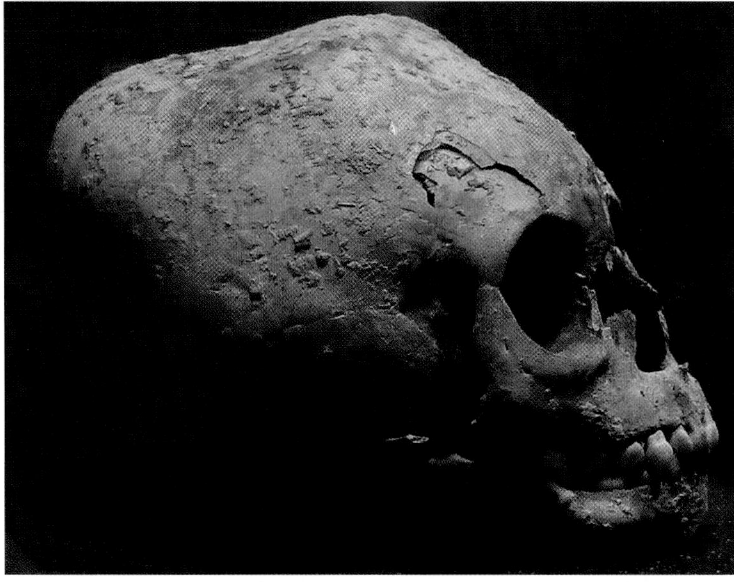

Above: A jade Olmec statuette showing an elongated head. *Below*: A Mayan skull with an elongated cranium now at the Archeology Museum in Mexico City.

Two photos taken in different years, of the display in the regional museum in Ica of some elongated skulls from Paracas.

Massive elongated skull from Paracas now at the museum in Ica.

Two amazing specimens of elongated skulls from Paracas, now at the museum in Ica.

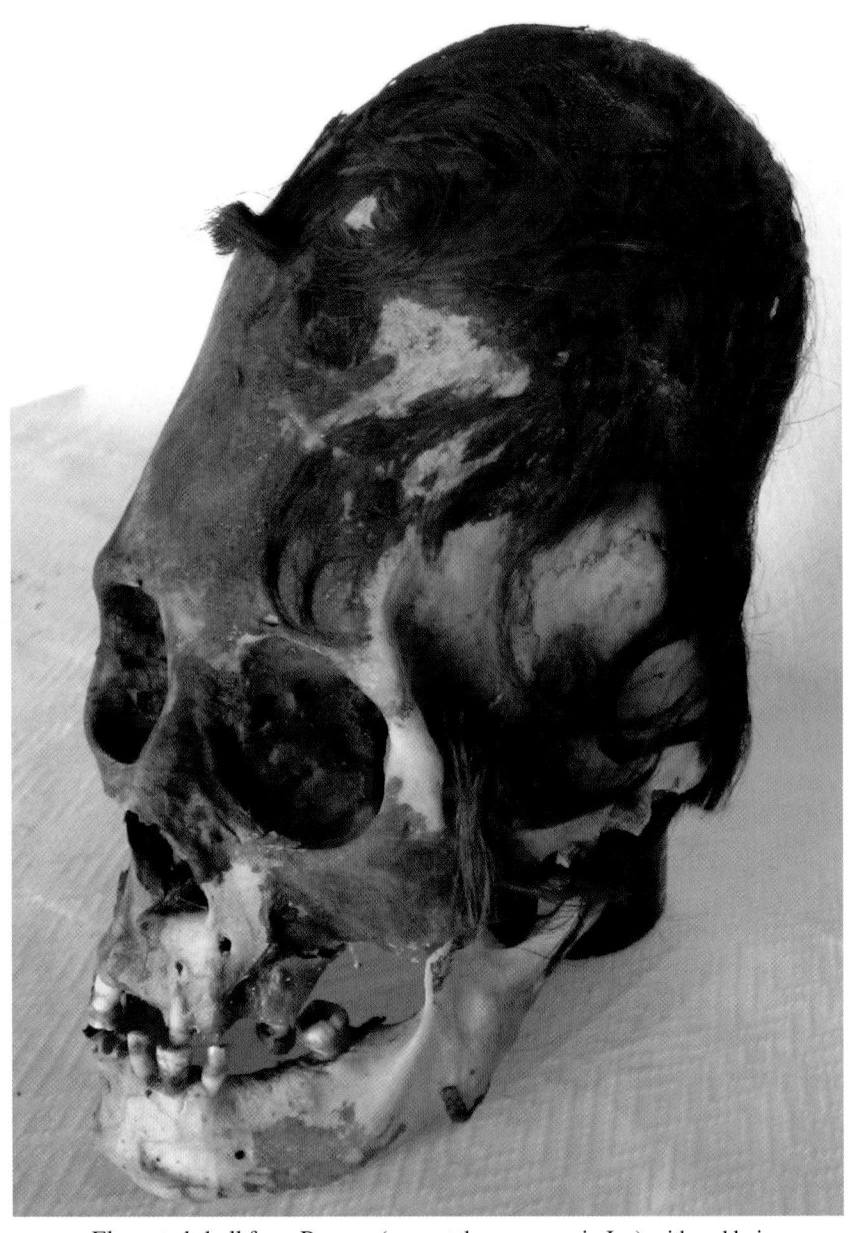

Elongated skull from Paracas (now at the museum in Ica) with red hair still attached to the cranium.

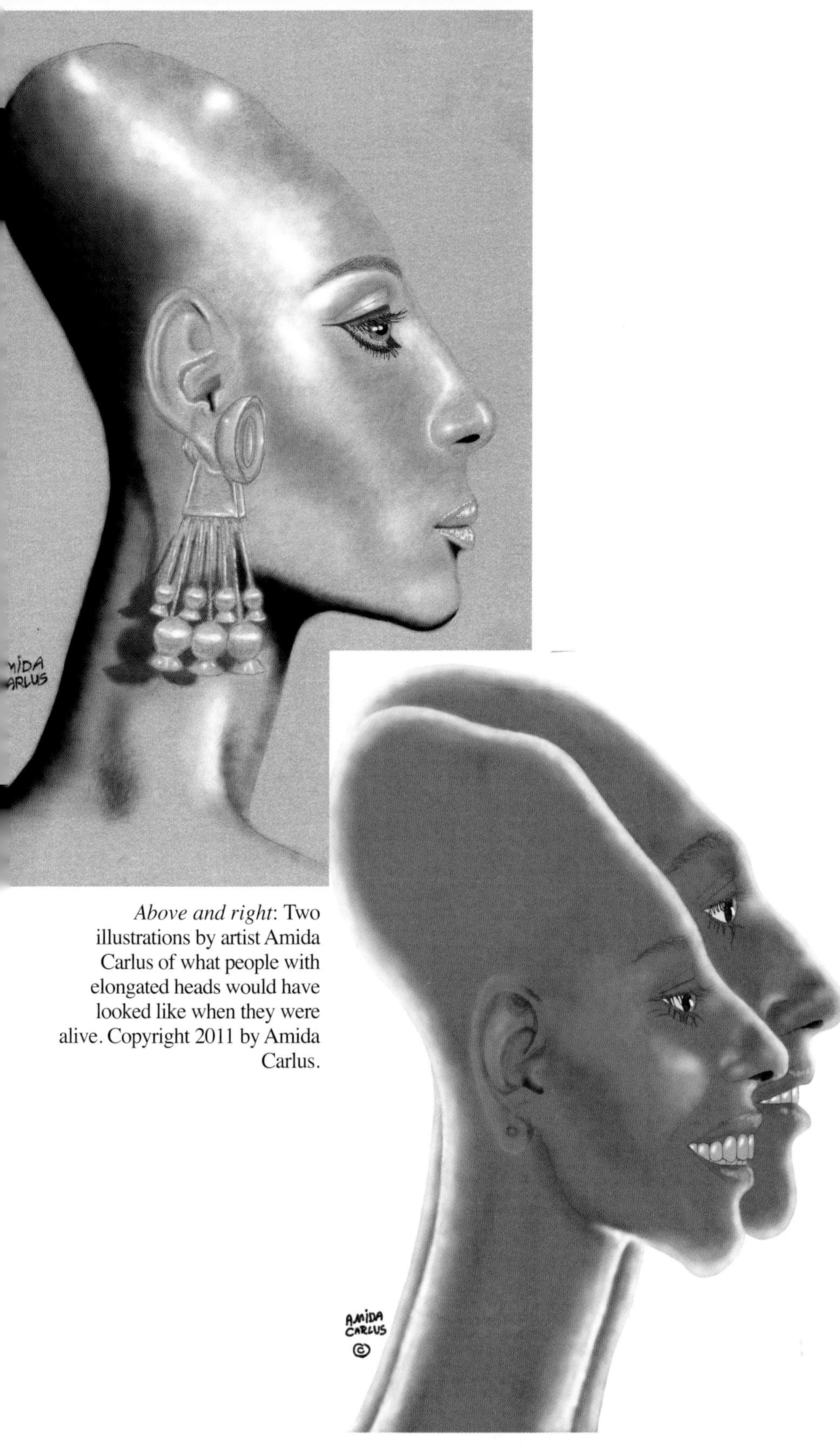

Above and right: Two illustrations by artist Amida Carlus of what people with elongated heads would have looked like when they were alive. Copyright 2011 by Amida Carlus.

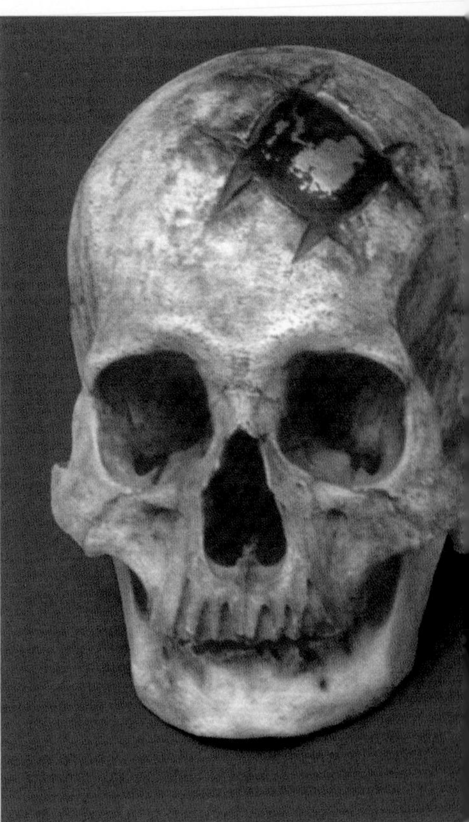

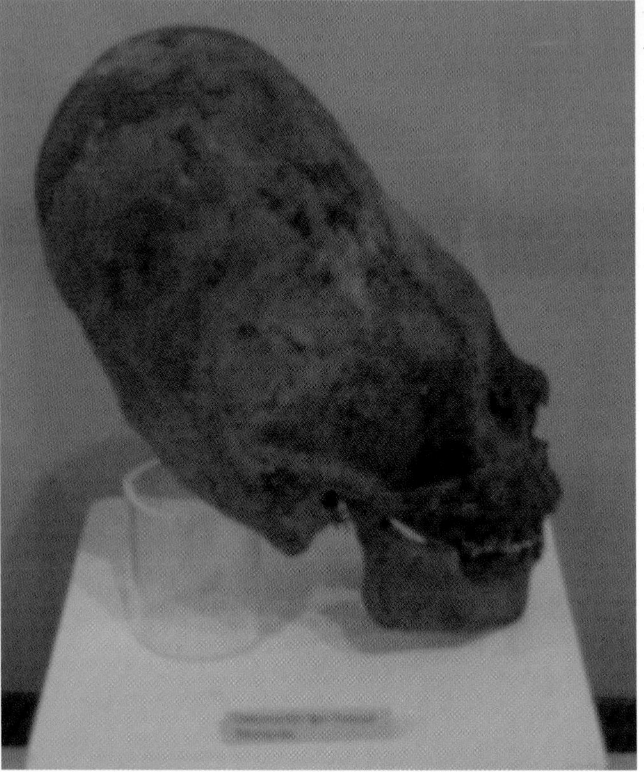

Above right: A trepanned skull v
gold plate inserted into the oper
now at the Gold Museum in Lir
Presumed to have come from cc
Peru around Paracas or Lima, d
unknown. *Above left and below*
Two different views of one of tl
Paracas skulls now at the Arche
Museum in Ica, Peru.

Above, right and below: Various elongated and otherwise deformed skulls, one in the rare "M" form (above), now at the Paracas History Museum.

Above and Below: Elongated and normal skulls once on display at the Tiwanaku Museum in Bolivia.

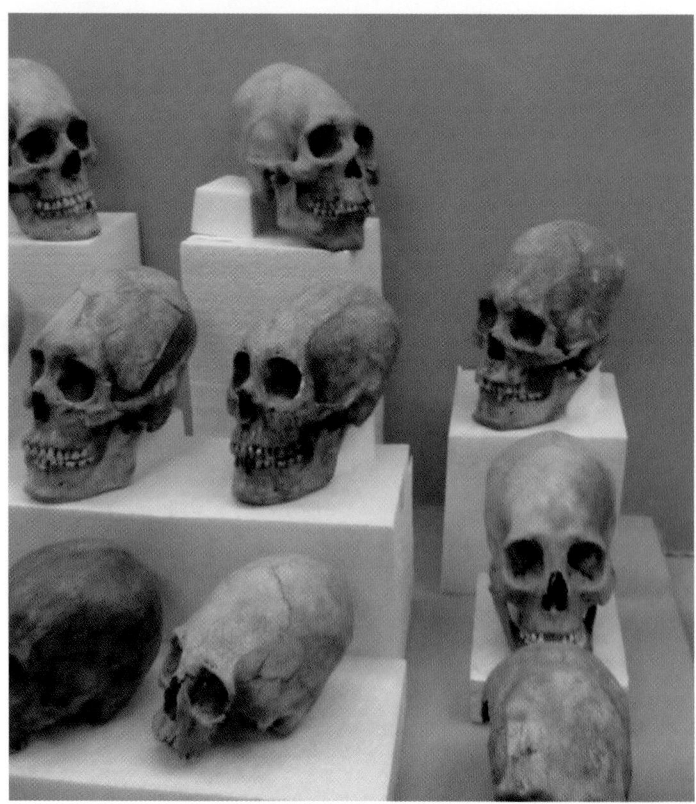

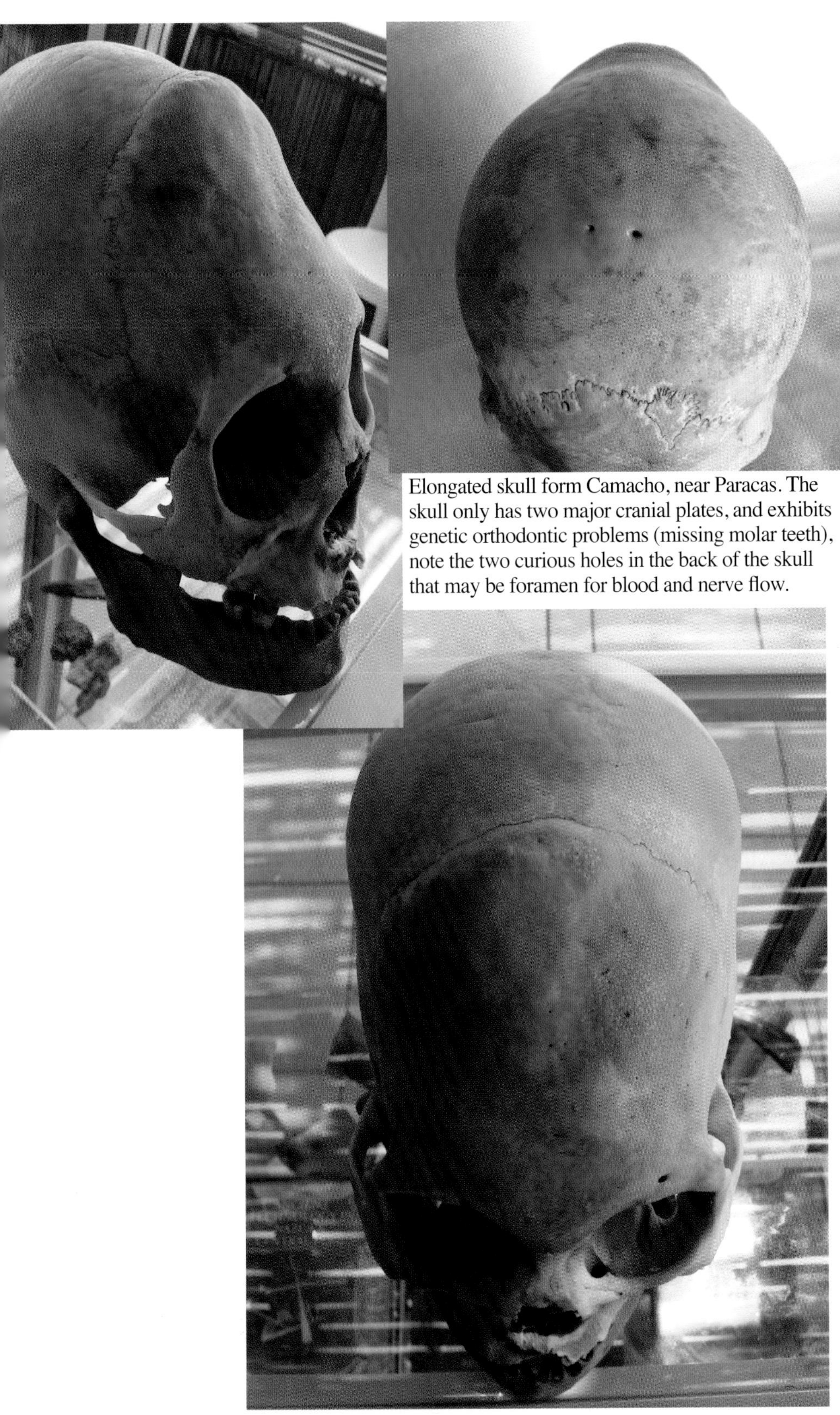

Elongated skull form Camacho, near Paracas. The skull only has two major cranial plates, and exhibits genetic orthodontic problems (missing molar teeth), note the two curious holes in the back of the skull that may be foramen for blood and nerve flow.

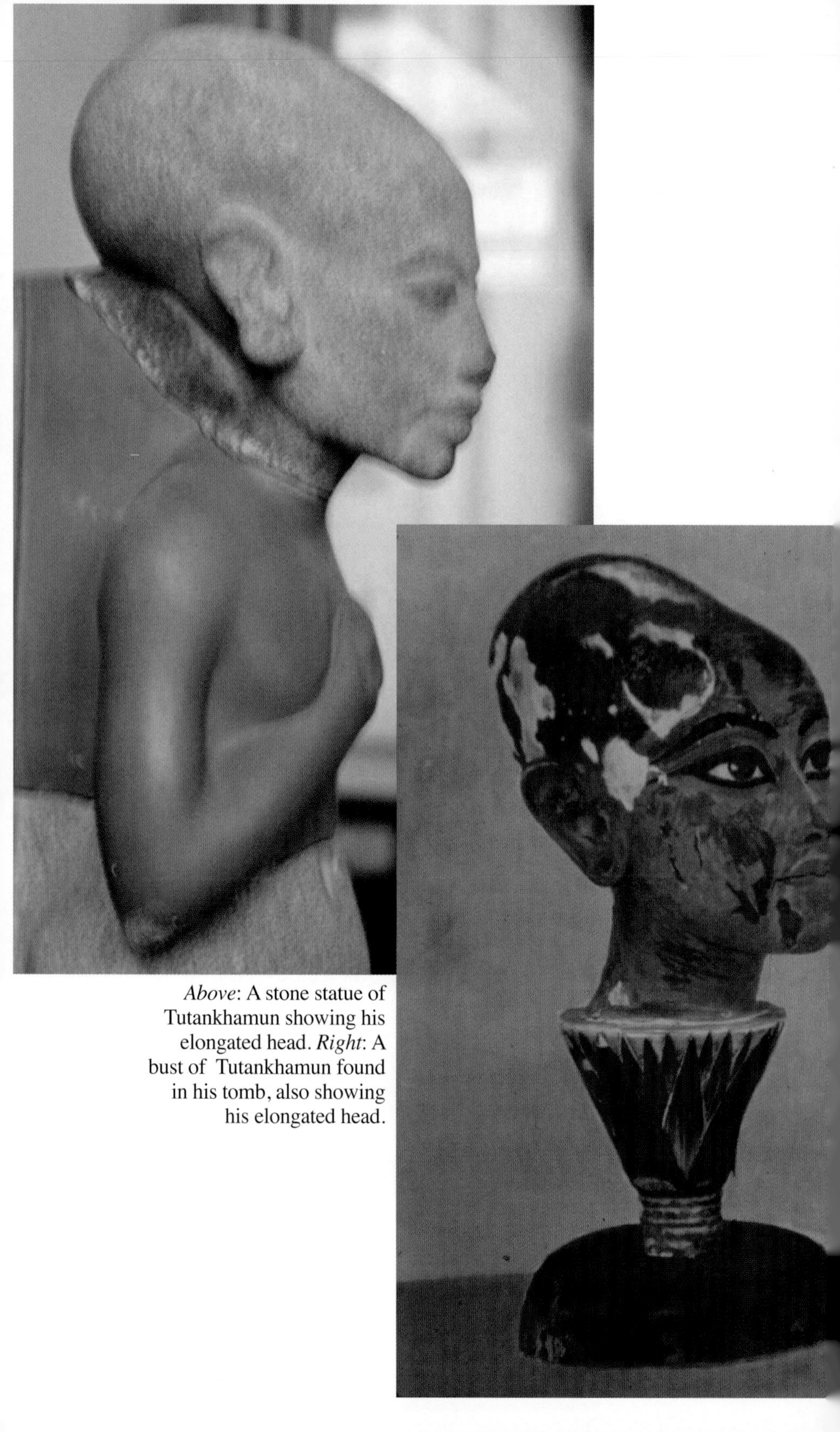

Above: A stone statue of Tutankhamun showing his elongated head. *Right*: A bust of Tutankhamun found in his tomb, also showing his elongated head.

Above and right: A bust of the Atonist princess Meketaton, now at the ro Museum, showing her xtremely elongated skull.

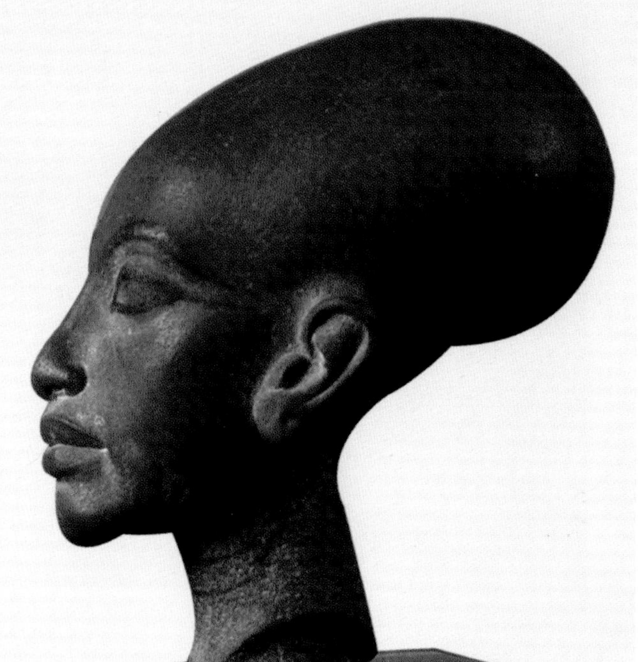

A Mangbetu mother and child, whose head is bound as an infant, circa 1930.

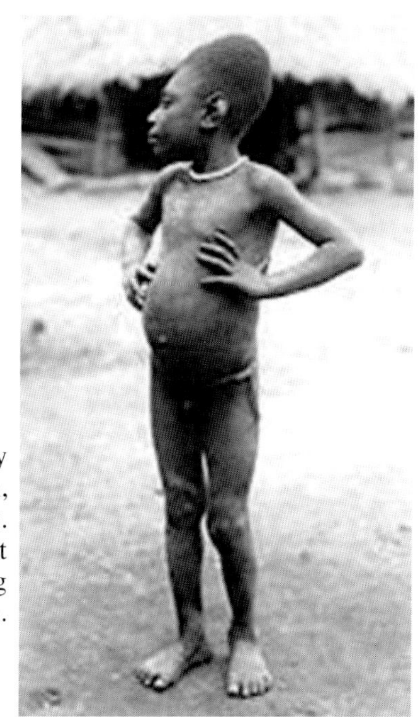

Right: A photo, circa 1940, of a young boy from Vanuatu with an extended cranium, known to be the result of headbinding. *Below*: A depiction of three of the Taoist immortals with the two on the right having the curious "M" cranium.

Above: A picture of a Flathead Indian of Montana painted in 1851. *Below*: One of the Taoist immortals depicted with an extremely long cranium.

CHAPTER 4

CRANIAL DEFORMATION IN SOUTH AMERICA

The world will never starve for want of wonders,
but for want of wonder.
—G.K. Chesterton

The Strange World of Paracas

Now that we have taken a close look at the Olmecs of Central America, let us turn to South America.

The oldest examples of cranial elongation and deformation we have from South America are from the Paracas people who lived on the coast of Peru, south of Lima, on the peninsula named after them. The elongated skulls from this culture are the ones most commonly cited in research papers to back up "alien-human hybrid" ideas and "descendants of Atlantis" theories. Photos of the skulls are prominent on search engines. But who were they really? What evidence is there in the real scientific journals?

The first archaeologist to study the Paracas culture was a Peruvian, Julio César Tello (April 11, 1880 – June 3, 1947) who was considered the "father of Peruvian archeology" and was America's first indigenous archaeologist. He not only made the major discoveries of the prehistoric Paracas culture, but also founded the National Museum of Archaeology and Anthropology in Lima.

The Wikipedia article on Tello reads as follows:

> Tello was born a "mountain Indian" in an Andean village in Huarochirí Province, Peru; his family spoke Quechua, the most widely spoken indigenous language in the nation. He was able to gain a first-class education

by persuading the Peruvian government to fund it. Tello completed his Bachelor's degree in medicine at the National University of San Marcos in Peru in 1909. While still a student, Tello studied the practice of trepanation among natives of Huarochirí and amassed a very large collection of skulls. He was also studying early pathologies in the population. His collection became the basis for a collection at his university. [The collection has since been moved, presumably, to the Museum of Archaeology and Anthropology in Lima, where it is in storage.[51] Even though San Marcos University can boast such illustrious alumni as Tello, it has no archaeological museum whatsoever, and its archaeology department is very poorly funded.]

He was awarded a scholarship by Harvard University, where he learned English and earned his Master's degree in anthropology in 1911. Next he went to Europe, where he studied archeology. In 1912 he attended the Congress of Americanists in England, a group in which he became prominent in later years. It was the beginning of his active international life.

Tello traveled widely during his career, and regularly invited other scholars to Peru, developing an international network. Although Tello published a number of papers in his lifetime, they appeared in little-known journals and newspapers, so they were not well-known then even to Spanish speakers.[52] For some time his findings and theories were not widely known outside of Peru because he did not publish in recognized academic journals.

In 1919 Tello was working with a team at the Chavín de Huantar archaeological site, where he discovered a stele since named for him, the Tello Obelisk. Construction of the first temple at this major religious center was dated to 850 BCE. The work of Tello and others established that the center had been a center of complex culture that lasted for several hundred years, to sometime between 500 and 300 BCE. Until late-20th century discoveries established the dates of the 5000-year-old Norte Chico site [now called

Caral], the Chavín culture was believed to be the oldest complex civilization in Peru.

Tello is best known for his discovery in 1927 of 429 mummy bundles in the Cerro Colorado area of Peru on the Paracas Peninsula. He first visited the site on July 26, 1925, following a trail that had begun in 1915 when he had purchased ancient textiles in Pisco. On 25 October 1927, Tello and his team uncovered the first of hundreds of ceremonial mummified bundle burials. He was the first in Peru to practice a scientific method of archaeological excavation, to preserve stratigraphy [the natural layering of organic materials over time] and elements to establish dating and context. In 1928 the team began to remove the mummies and textiles for safekeeping.[51] His findings and interpretations have been the most significant source of information regarding the Paracas culture, which dates to 750 BCE—100 CE.

Sadly, few archaeologists have conducted much in the way of study of the Paracas culture since Tello's time, and the mummy bundles, each containing a full skeleton with most likely elongated skull, languish in storage in Lima museums. The two volume work, *Paracas: Cavernas Y Necropolis*, by Julio C. Tello and T. Mejia Xesspe was published by the Universidad Nacional Mayor de San Marcos of Lima in 1979.[53] The entire Paracas Peninsula is now a national park/wildlife refuge, so no further archaeological digs will be likely in the foreseeable future. So this leaves us with Tello's dates for the dawn and dusk of the Paracas...

Or does it? This is where we hit somewhat of a murky area. Tello's whole theory about the age of the Paracas culture was mainly based on his seeing similarities between Chavin culture ceramics and some of those found at Paracas. As one archaeologist, Anna McBain stated

Julio Tello, circa 1940.

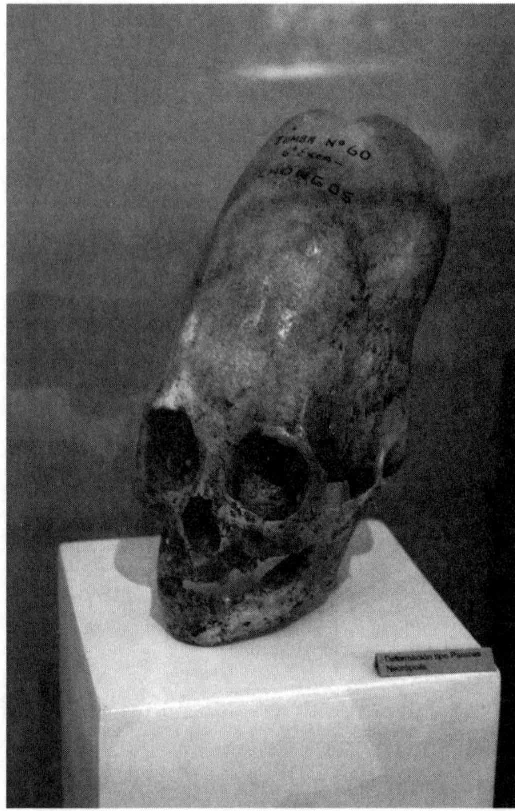

Massive elongated skull from Paracas.

when this was revealed to her: "perhaps the Paracas people simply traded with the Chavin..."[54] This is indeed quite plausible; just because ceramics depicting similar art styles and animal characters were shared by the Paracas and Chavin cultures, it does not mean that the Chavin were the source of the Paracas people. Their presence on the coast of Peru has not been fully documented due to the lack of research, and the fact that any research is done is almost totally dependent on the work of Tello.

What dating techniques would have been available to Tello in 1928? Carbon 14 dating wasn't available until the late 1940s when Dr. Willard F. Libby invented it, for which he received the Nobel Prize in chemistry in 1960.[55] Because Chavin was the oldest culture known at the time, Tello must have assumed that any other culture, including the Paracas, was more recent.

What adds to the lack of logic of this assumption is that Chavin developed in the highlands of Peru, while the Paracas were a coastal people. Does it not make sense that the Paracas people arrived by sea rather than descended from the highlands? No elongated skulls have been found at Chavin, while more have been discovered in Paracas than anywhere in Peru, and indeed the world. Even if the Chavin culture had an influence on Paracas, they clearly would have pre-dated such an influence, but by how far back in time, no one knows.

Such a phenomenon as cranial deformation would hardly appear out of thin air, because it was a very specific and complicated process, one not to be taken lightly. In the case of the Paracas people, the actual implements used to achieve the binding included a sling-type cord; it was doubled across the front of the skull of a child, presumably a few months old, with a padded backing that fitted to the back of the head. A wooden stick was supposedly placed inside the backing for rigidity, and the cords were gradually tightened, over months and years, in order to alter the shape of the cranium.[56]

An example of this apparatus, in perfect condition, is in the Paracas History Museum, made of local cotton, which is indigenous to the area. The resulting cranial manipulation results (as seen in many of the elongated skulls in the museum), in the presence of a pronounced bump in the upper forehead of each individual, in the area where the two cords were not present.

However, what is even more intriguing, is that the Paracas History Museum also has examples of skulls devoid of such a bump, with the forehead gently curving back to the apex of the head. From this one could possibly presume that another technique was used, perhaps with a curved board placed on the forehead, with cords wrapped around it and gradually tightened. Or there is the more amazing possibility: that some of the Paracas people in fact had naturally elongated skulls. In other words, they were born that way!

An amazingly detailed sketch, made by Johan Jacob von Tschudi, MD and presented in his book, co-authored by Mariano E. Rivero, published in 1851, *Peruvian Antiquities*[57] shows a human fetus (supposedly Inca but of no detailed documented source or location), with a huge elongated skull. Dr. von Tschudi made the sketch from a specimen he had in his collection, of which he states "the same formation of the head presents itself in children yet unborn; and of this truth we have had convincing proof in sight of a fetus enclosed in the womb of a mummy of a pregnant woman... which is, at this moment, in our private collection."[57] He estimated the fetus to be seven months old, and thus fully physically formed.

What has happened to this mummy of mother and baby is unknown, but it is presumed, along with so many of the amazing human artifacts of ancient Peru, to be somewhere in storage, at or near one of the Lima museums.

The "fetus" from von Tschudi, 1851.

Dr. von Tschudi was a "doctor in philosophy, medicine, and surgery etc., etc., and a member of various societies of medicine" which were credentials crucial to what he reports: "The singular conformation of the Peruvian crania" found in what appeared to be two kinds of the "three distinct races (who) dwelt there before the foundation of the kingdom of the Incas."[57] The Inca are believed to have come from the area of Lake Titicaca, and more specifically the Island of the Sun (according to legend), now in present-day Bolivia (and/or the enigmatic place known as Tiwanaku (Tiahuanaco) also located in Bolivia) about the year 900 AD.

Von Tschudi's account raises the possibility that at least some of the cranial elongation may have been the result, in some cases, of natural genetics. The Paracas History Museum acquired, as of September 2011, the skull of a baby, less than a year old, which seems to be as cranially deformed as that of two other skulls found in the same tomb in the area called Chongos, 20 minutes drive from the Paracas town of Chaco. Chongos was a major burial site, not only for the Paracas culture, but also for those that came after them: the Nazca, Wari (Huari), Chincha and Inca. The baby's head, along with that of a female, guessed to be approximately 20 years old, was found wrapped in ceremonial cotton cloth, and this has not been removed as of this writing, due to the uniqueness of the specimen. Chongos is one of four burial grounds in the immediate Paracas Peninsula area where elongated skulls have been found in profusion, without the presence of many "normal" skulls, at least from the Paracas period. This clearly indicates that these represent people of a special social class, a point we will examine more fully later.[56]

A great physical presence in the Paracas Peninsula, one shown to hundreds of tourists from boats each day, is the Candelabra. This geoglyph, which of course reminds one immediately of the more famous Nazca lines to the east, is believed by most researchers and local tour guides to represent a large cactus, presumably made by the Paracas people due to a hallucinogenic fetish. Such a claim is based on nothing but pure speculation. The glyph is almost 600 feet in height and forms a depression almost a meter deep in hard quartz crystal and salt encrusted sand. It is believed to be contemporary with the Paracas culture, about 2,000 to 3,000 years old, but it cannot be independently dated.[60]

Due to its location right by the ocean, and being at a 20 degree angle incline in relation to the ocean surface, it could very well have been a navigational marker for seafarers and would have been highly useful as such. Especially during later, brighter phases of the moon, and considering that Paracas is rarely if ever overcast with clouds, such a large symbol would be seen for several miles out to sea, even at night. But what is it pointing to? It's length points almost due south, and the nearest landmass in that direction is Antarctica! The author and professor Charles Hapgood theorized that Antarctica was Atlantis, but that is taking us off track here![58]

The local tour guides tell their visitors that the Paracas people may well have been the makers of the Candelabra, and that it represents a giant cactus, more specifically the San Pedro (named after St. Peter of course) which has grown in the area since time immemorial. It has been an incredibly useful plant for thousands of years with the spines used for knitting and possibly fish hook barbs, the fiber for clothing, and the mescaline essence in its core for shamanic practices.[56] However, to describe the builders as being worshippers of a so-called "cactus cult" is to debase them and their history. This is, unfortunately, a very common practice in Peru, especially in the tourism field. Those who do not have sufficient information about ancient people and places ascribe all kinds of behaviors to cult activity; there were apparently water cults, cactus cults, feline cults, etc. Such simple labels say absolutely nothing.

So, at Paracas we have an ancient people living by the ocean, next to a huge geoglyph, in the shape of a trident—the international symbol of seapeople—which could only have been made by an engineer/designer on some sort of sea craft instructing a presumably large group of individuals on shore about how to make it. Closer examination of the area, especially the northern portion of the peninsula where the town of Chaco is located, shows that it is a very sheltered, and quite massive bay. The prevailing winds come from the south, and the headlands, separating the northern and southern parts of the peninsula, make the upper bay sheltered from the creation of large waves. Thus, it was and is a perfect harbor. The presence of a large fishing fleet there today indicates that this natural harbor is still in use.

People who live by the ocean tend to be fisherman, and this is and has always been the case in Paracas. The first inhabitants, a

presumed Stone Age culture, are believed to have lived in the area at least 10,000 years ago. The Paracas History Museum has a large collection of implements that they left behind, mainly made from local volcanic rocks, such as axes, spear points, knives, etc. The dating is speculative; no scientific physical or chemical tests have been conducted, so the age may be more recent, and yet could very well be far more distant.

In the archaeological record left behind by the actual Paracas culture, which can be identified by the existence of their very specific ceramic styles mixed in with edible mollusks and fish bones, a huge array of marine species can be found; and not just fish, but also sea lion bones and teeth, as well as the remains of dolphin and porpoise skeletons. The Paracas History Museum has a well-preserved pair of sandals, made from sea lion skin, prominently on display.

What this tells us is that the Paracas were not simple fishermen, catching their prey using simple tools in shallow waters. They must have had boats, or even ships large enough to exploit sea mammals, whose intelligence and speed would require sophisticated sea faring abilities. And so what materials would have been present to make these water craft from? There are no trees present today, of a size that would easily allow for the construction of boats and the area has been stark desert for thousands of years.

What we do find, however, are totora reeds, in great abundance, growing thickly all along the coastline where there is fresh water. Totora is indigenous to Lake Titicaca, some 700 km to the southeast of Paracas, and has been used for thousands of years for the roofing of houses, mats, and of course, boats. The Titicaca craft, called Tora in the Quechua language, are incredibly sturdy and stiff boats and as buoyant as cork. Thor Heyerdahl, the intrepid Norwegian explorer best known for his Kon Tiki expedition from Peru to Easter Island in 1947, flew a master Tora maker to Egypt in 1970 to assist in the construction of the Ra reed ship, made from local papyrus; it is with this, and the Ra II that he endeavored to travel from North Africa to the Americas.[62]

Therefore, we can easily presume that the Paracas used totora as a boat/ship building material, but since it is a natural material easily degraded by the sun and other natural elements over time, nothing of their existence would appear in the archaeological record. What is also significant is that whale bones have been found in the archaeological record of the Paracas culture, and more specifically sperm whale bones.[56] These animals are very intelligent, and have been historically difficult to pursue, let alone harvest.

This suggests that the Paracas people were open ocean sailors, and may have incorporated sails made of local cotton with masts and booms of native bamboo to make the hunting of such large prey possible. It seems that the Paracas people were not primitive people living along the coastline, but sophisticated masters of the ocean, as well as agricultural and hydrological engineers, which we shall get into later.

The mollusk known as the spiny oyster, *Spondylus princeps,* is only found in the waters off of Ecuador far to the north of Peru. The presence of jewelry adornments made from its shell in the graves of the Paracas, specifically in the tombs with those with elongated skulls, tells us that either the people of Ecuador came south, presumably by ship, to trade with the Paracas, or that the latter sailed north in search of this highly prized shell.[64]

The finest textiles ever made in what we presently call Peru are from Paracas… this from an "early" and "primitive" culture? The burial garments, called mantles, are highly prized by international collectors, some fetching prices of up to and beyond a million dollars. Part of the reason why Tello was attracted to the Paracas area was the flaunting, on the local markets, of these ceremonial garments. There fineness of weave and incredibly rich colors ensured that they found their way into private Peruvian and foreign collections, as well as international museums, as far back as the early 1900s.

These mantles were worn exclusively by the highest of the Paracas elite, and were buried with them. The fact that the majority of these ritual garments were found in graves of individuals with elongated skulls tells us that the elongation was clearly the domain and characteristic solely of the royal and priestly classes. Since in most Andean cultures the two were the same, as in the case of the

Inca, where the eldest son was the high Inca (king) and the second born the high priest, it was the outward symbol of distinction.[65]

Perhaps what is most striking about the mantles, aside from the fineness of their execution and bright colors, is the symbolism involved: men with large smiling snakes coming out of their mouths, holding decapitated heads, with multiple strange, almost UFO shaped bird-like images crowning their heads. The faces of these "men" almost invariably have cat-like whiskers, with faces, and cat-like teeth in their mouths.[59] It was this seeming obsession with the feline that caused Julio Tello to surmise that the Paracas were influenced by the Chavin, and thus allowed him to stamp the dates of this culture's existence so hard that they have stuck for more than 80 years until being pried open by new research. Tello's presumption that the Paracas culture began as an offshoot of the Chavin must be examined in a new light. Only extensive and proper scientific dating and future excavations will allow us to properly figure out where the Paracas people came from, and from where the cranial deformation process originated.

Very little has been studied about the Paracas people since Tello's time. His archaeological digs at Cabeza Larga (Elongated Head) and Cerro Colorado (Red Hill) which were funeral areas, as well as Chongos and others, have been ignored since the 1930s, and thus have filled in with sand and been largely forgotten. What is also intriguing about these people, apart from the abundance of elongated skulls found which we will get into soon, are their burial practices and the houses in which they lived.

The only area where Paracas remains have been found on the peninsula proper is at the northern end, at the large so called Necropolis site. This area was where the Paracas people not only buried their dead, but also lived. Local historian, archaeologist, and museum owner Juan Navarro believes the "village" site was two kilometers long but has no idea as to the population size; however, the close proximity of one "sunken" house to the next would suggest several thousand.

The people decided to live underground, in homes that could accommodate an entire family. The reasoning behind this is vague, but was probably due to the lack of stone or wood for building

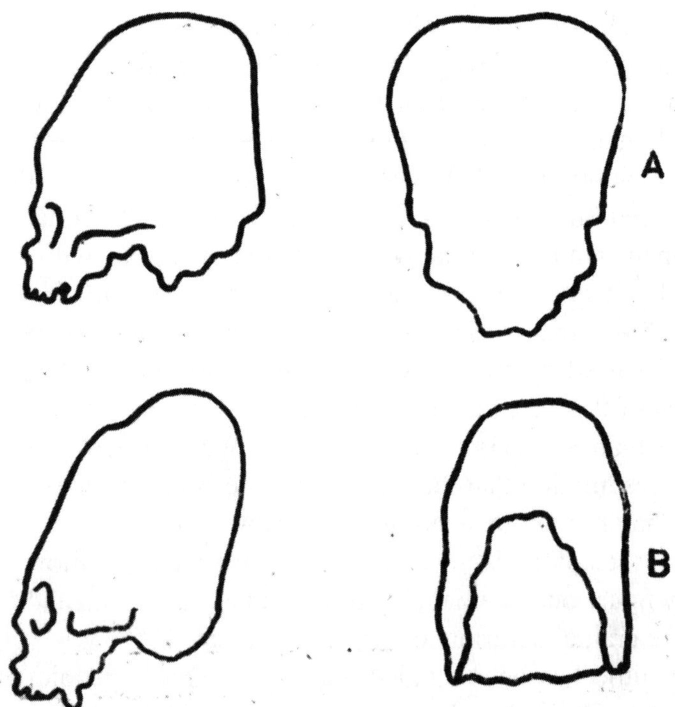

Early drawing of skulls from Paracas by Julio Tello, circa 1940.

above ground dwellings, and the fact that there is a constant strong wind that blows in off the coast. The word Paracas, which is a Spanish term, is based on the Inca Quechua words Para Aco, meaning "raining sand." The term was most likely coined by the Inca, who were the last native inhabitants in the 15th and 16th centuries AD; what the "Paracas" in fact called themselves, and what language they spoke is unclear.

Wave after wave of cultures moved in and occupied the area. After (and during) the Paracas came the Nazca, a more warlike people who inhabited the area from 200 BC to 700 AD; Wari (Huari) 600 to 1000 AD; Chincha 1100 to 1470 AD; and finally the Inca, from 1470 to 1532 AD.[66] Such a succession and overlap of people from different places would have watered down any cultural and linguistic expressions of the Paracas, and thus their graves are what hold the keys to our present understanding of them.

Their graves are intriguing. Both at Cabeza Larga and Cerro

Colorado we find that graves were not occupied by a single individual, but by groups of up to 40, according to Tello. These were presumably family mausoleums. A trap door on the surface, which could be opened when a family member was to be interred, led to a family plot shaped like an inverted mushroom. Sufficient space was available, as has been said, to house multiple bodies. The way in which they were wrapped is also intriguing. These royal personages were placed in the tombs with multiple layers of mantles, at least three wrapped around them, and that is why Tello was able to acquire such a wealth of them.

His departure from the area, supposedly about 1930, due to internal upheavals in the Peruvian federal government leading to insurrections and the cutting of financial support to him and other scholars, is what caused the Paracas archaeological studies to be abandoned. Tomb robbers, called huaqeros in Spanish (from huaca, which is Quechua for sacred), soon moved in and plundered all of the burial and ceremonial areas of the Paracas and later cultures of the area. In fact, the great collections of Paracas cultural artifacts found today in American and European museums were acquired on the black market, mainly in the 1930s, by purchasing directly or indirectly from huaqeros.

The royal funeral bundles look, in cross section, like eggs, and some have inferred that this practice, along with the existence of food and ceremonial paraphernalia in the tombs, including feathers from the Amazon jungle, meant that the Paracas believed in an afterlife. The tombs of the common people, which clearly outnumbered the royal class, consist mainly of bodies poorly wrapped in simple cotton shrouds. So common and spread out are these "worker tombs" along the seashore and inland amongst agricultural areas, that it is clear a substantial population provided for and most likely worked for the royal elongated skull class.

All of this is meant to show that these Paracas people were "different" in many ways from what, according to Tello, may have been their contemporaries in the highlands of Peru. They were master farmers; what is intriguing about the coastal area where they lived is that the water table is slightly below the soil even to this day. The water itself comes from the higher mountains to the east, and perhaps as far as the Andes, and percolates its way down just below the soil. This means that the Paracas, who lived in a very dry environment, indeed so dry that it only rains about one inch per year, never had to irrigate their crops, although they did have to dig fields lower in some areas to stay close to the water level.

And the area that the Paracas in fact inhabited is actually quite immense. The Paracas were masters of agriculture and water diversion. From Chincha, about 50 km north of Paracas to Huaca Alverado and Huaca Soto, and southwest by more than 200 km into what is now called Nazca. Chincha is home to a cluster of quite immense adobe pyramid structures purported to have been made by the Paracas people.[66]

There is a lot of evidence that the entire Peruvian, and perhaps northern Chilean coast, underwent a massive climate shift in about the 5th century AD. Cultures such as the Moche, who were the builders of massive adobe pyramids in the northern part of present day Peru near Tucume, died out at this time due to drought. So the area of Paracas, and inland into what is now called the Nazca area, may have been far more agriculturally productive in the past. Combine this fact with the ingenious ways in which they lived and prepared their burials, and you are clearly not looking at a "primitive" hunter-gatherer people, but perhaps one that chose this area in which to live, and perhaps were not from the area at all originally.

If you use a search engine such as Google and type in "elongated heads," or "elongated skulls," even if you type in Egyptian elongated skulls, invariably photos show up, in great numbers, of Paracas ones. Why? Because more skulls have been found here of an elongated nature than anywhere else on earth. This is partly because of the extremely dry nature of the climate, excellently preserving skeletal material even if it is thousands of years old.

The largest elongated skulls on earth are all on display in the regional museum of Ica, which is about an hour's drive from Paracas, and all were found at the same site very close to the small town of Chaco on the Paracas Peninsula. This was "cone head central!" It is the previously mentioned ancient ceremonial and burialsite called Chongos, now on private property and completely off limits to public or archaeological team visits. It stretches for more than a mile, and hosts six large but crumbling adobe pyramid-like buildings.

Sr. Juan Navarro of the Paracas History Museum suggests that these pyramids date from the Wari (Huari) period of approximately 600 to 1000 AD, but the Wari are not well known for their use of adobe, whereas the Paracas are. Scattered across the desert sands from temples 1 to 6, are thousands of pottery shards, those from the Inca period being the most recent The shards display mainly geometric designs in limited colors, and date back through the Chincha, Wari, Nazca and then Paracas cultures. The pottery of the Paracas is most commonly glazed with simple black burnished oxides, sometimes with hints of lighter tones in the shapes of animals, the snake being the most common symbol.

Two other funerary sites mentioned above are Cerro Colorado and Cabeza Larga. Cerro Colorado is regarded as being the older of the two by Tello and his present day counterpart Juan Navarro, and is separated from Cabeza Larga by about one quarter kilometer. In fact, a modern road bisects these two sites, both of which are inside the Paracas marine preserve. On the site of Cabeza Larga once stood the Julio Tello Museum, which was badly damaged after the 2007 earthquake that devastated the whole area, especially the town of Pisco to the north.

The entire contents of the museum, including many specimens of elongated skulls, were removed to the regional museum at Ica, and also the University of San Marcos in Lima, Tello's alma mater. That collection has since gone to the Museum of Archaeology and Anthropology in the Pueblo Libre neighborhood of Lima, which Tello founded. Unfortunately, much of this collection is in storage, not on display.

At present, only two elongated skulls are on display at this museum, along with 3 more normal shaped trepanned skulls. Why? That is a great question for Sherlock Holmes. The main reason given is that the museum likes to constantly "turn over" its displays, and it has such a huge inventory of artifacts that this may well be the case… or is it?

Perhaps there are more nefarious reasons afoot. Is it possible that the museum, and perhaps governing bodies, have been told not to display these skulls? Are they so odd—possibly representing a subspecies of human—that the "powers that be" don't want the general public to know of their existence?

The regional museum in Ica has dozens, and perhaps as many as one hundred of these elongated skulls, with only six or so on display. The others are kept in simple cardboard boxes, some wrapped in paper and some not. They are not on display supposedly due to budget considerations, but who knows the full truth on this?

The Paracas, as I have said, covered much more area than just the peninsula named after them. In fact, the area around Nazca, home of the famous lines was also where they lived, prior to the Nazca. It is believed, according to conventional archaeology, that the Paracas died out, or were absorbed into the Nazca, about 200 BC. The presence of Nazca pottery in abundance around Paracas shows that they certainly lived there at some time.

So called "trophy heads" of the Nazca (supposedly skulls worn around the waist of warriors to indicate their prowess) include examples of the elongated skulls of the Paracas people, and the theory was that the Paracas were subdued by the Nazca. The process of elongation slowly died out, while the Paracas art of trepanation was adopted by the Nazca. However, new evidence suggests that the "trophy heads" were actually sacrifices made to the rain god or gods of the Nazca.

Does this mean that the most prominent members of the society could have been sacrificed, or even volunteered to the rain deities in order to help the crops grow? It does seem possible. The reasons for the demise of the Nazca were rain related—specifically lack of rain. The area, like Paracas, receives an inch or less of rain per year, but the water table is lower. This is why elaborate water systems whose source was the distant eastern mountains were developed.

These water systems, as well as the lines of Nazca are purported to have been made by the Nazca people, but for this

The Candelabra of the Andes at Paracas seen from the ocean.

there is no proof. All that is known about these people is that they were master potters and had an agrarian society. The great temple complex of Cahuachi, located about 30 km from what is now the city of Nazca, is said to have been built between 500 BC and 500 AD, and covering a remarkable 24 square kilometers. Paracas pottery has been found in the area, so they may have been the first of the builders. The builders of the aquaduct systems may also have been the Paracas.

The famous Nazca lines are now known to have been made over two distinct periods; the animal shapes were made first, and the lines and trapezoidal features later. The simple proof of this is that the lines, in many cases, cut straight through the animal figures; this shows that the animal shapes were made first.

As has already been said, a massive El Nino event in about 500

AD seems to have brought about the downfall of the Nazca culture, but this was not the only factor in their demise. Recent research shows that the Nazca people had progressively been denuding the natural forests of the area, presumably for firewood and to increase the amount of arable land for their growing population. What they failed to realize was that the roots of the trees held the water and scant soil in check; the result of their removal was starvation, and a population implosion.

As has been remarked, the Nazca did adopt trepanation from the Paracas, but not cranial elongation. And the actual reasoning behind trepanation, which was very common amongst the Paracas, is as yet unclear.

Warfare would no doubt have resulted in many cerebral contusions, and thus surgery of this nature would assist in releasing pressure from a buildup of fluid between the brain and the interior of the skull. Tumors, of course, could also be removed if a certain area of the skull cap was removed in order to obtain access. And, there could have been a spiritual component to it—evil spirits being released by opening up the skull.

What is most intriguing is not the number of trepanations that were performed, but the number of patients that seemingly survived. For example, in October 2011 the Paracas History Museum received an elongated skull with signs of a trepanation having been performed on the left side of the crown of the head. The original circular hole was approximately 9 cm in diameter, and completely healed over!

Sr. Juan Navarro has found thin sheets of silver and gold at the grave sites where trepanned elongated skulls have been found, the sheets matching the size of the trepanations. This tells us that these people may very well have used silver and gold as coverings for the open areas of the skulls, protecting the exposed areas from infection. Gold is used in modern surgery, along with stainless steel, because the body does not react to its presence, while silver has been known for centuries (if not millennia) for its antibacterial properties.

With regard to the elongated skulls, what of the possibility that some of the skulls were at least not cranially deformed, at least not completely—that there is a genetic component involved?

Three skulls presently on display at the Ica Regional Museum, taken from the burial area of Chongos, are not simply deformed and elongated, they seem to be larger in volume than most of the other "coneheads" found in the world. Some have speculated, though not proven, that their internal cranial capacity is between 2.0 and 2.4 liters. The average modern human skull averages about 1.3 liters, so this is truly curious.

Also, many of the skulls at the Paracas History Museum have auburn, wavy hair. Native peoples of North, Central and South America are predominantly known to have almost jet black hair. Is the reddish color present in the Paracas skulls, as well as others found in graves at Nazca, the result of sun exposure? Is it natural dye applied while the individual was still alive, like a henna treatment? Or could it be a genetic characteristic that comes from ancestry as yet unidentified, and foreign to the Americas?

Oral traditions amongst the Maori of New Zealand (Aotearoa) and observations of the first Europeans to Easter Island (especially Jakob Roggeveen, who named it on Easter Sunday 1722) speak of the existence of pre-Polynesian people inhabiting these areas and having reddish or even blonde hair.[68]

Two other characteristics worth noting are the lack of molars in many of the elongated skulls in the Paracas History Museum and strange pencil-tip-size holes in the back of the head. Numerous medical doctors have visited the museum, not through specific invitation but because they happened to be in Paracas on holiday. All remarked that the lack of molars could have been a genetic trait, and not likely the result of a poor diet. The Paracas area was rich in corn, beans, tubers, peanuts, fish and other food stuffs, so it is doubtful that their diet lacked nutrients. But, this could be a genetic trait resulting from the royal class inbreeding. A lack of molars has also been observed in elongated skulls found in Iraq.

The holes are even more intriguing. It is normal for humans to have two holes in the lower jaw, one on each side; these are the mental foramen, holes where nerves and blood vessels exit from the interior of the jaw to assist in sensation and blood flow to this part of the head. But in the elongated skulls, two similar holes are found at the back of the skull, just below the crown. According to the medical personnel who witnessed these in person at the Paracas History Museum, they should not be there; but the doctors did agree that their presence suggests that nature may have provided these (through natural evolution?), to help feed the elongated part of the skull.

Finally, and most bewildering, are the cranial plates. In the average human skull, there are three main plates which make up the upper portion of the cranial area, the frontal plate, and behind it two parietal plates. These three are divided by what are called sutures; joints which allow a youth's skull to form, and fuse together over time.

The intriguing thing is that some of the elongated skulls from Paracas, especially the three on display at the Ica Regional Museum, show only the presence of two major plates, one frontal and one parietal. The medical professionals who have observed the examples at the Paracas History Museum were completely baffled; they offer suggestions, but none that they believed truly explained this phenomenon.

So is this the end of our story? Far from it! From here we go east, to Lake Titicaca, and pick up the thread again, this time with the people of Tiwanaku, or Tiahuanco if you prefer.

Tiwanaku the Mysterious

When one thinks of strange and mysterious places, Tiwanaku could clearly top the list. Tiwanaku sits, a forlorn group of five major buildings in various stages of reconstruction, about 20 kilometers south of the edge of Lake Titicaca. The site was first recorded in written history by Spanish conquistador and self-acclaimed 'first chronicler of the Indies' Pedro Cieza de León. Leon stumbled upon the remains of Tiwanaku in 1549 while searching for the Inca capital Collasuyu. Some have hypothesized that Tiwanaku's modern name is related to the Aymara term taypiqala, meaning "stone in the center," alluding to the belief that it lay at the center of the world.[69] However, the name by which Tiwanaku was known to its inhabitants has been lost, as the people of Tiwanaku apparently had no written language.

Its population is believed to have been at least 100,000 people, and some estimate even as high as 1,000,000 at its peak; but where are the architectural remains of such a huge population? It is most probable that adobe, or hardened and formed mud bricks were the most common material for walls, as is evident in the traditional construction techniques observed today. Roofs would have been made out of grass or straw. These are of course organic materials, and it would not take much time, once abandoned, for them to return to the soil, leaving no trace. And there is evidence that the buildings were abandoned as the population would have plummeted in the area due to the poor weather. The temple structures were made of stone and thus were more durable, and it appears that they stood until the last of the culture, the first Inca, left around the year 900 AD. Later people may have quite quickly commenced deconstructing the stone buildings to use as material for their own structures.

The area around Tiwanaku may have been inhabited as early as 1500 BC (or earlier) as a small agriculturally-based village. Most research, though, is based around the Tiwanaku IV and V periods between AD 300 and AD 1000, during which Tiwanaku grew significantly in power. During the time period between 300 BC and AD 300 Tiwanaku is thought to have been a moral and cosmological center to which many people made pilgrimages.[70] Its demise, as noted above, was about 900 AD as the result of a 40 year long El Nino induced drought that made living in this tough land at 14,000 feet elevation next to impossible.

Arthur Posnansky, who spent 50 years of his life studying this site declared it to be as old as 15,000 years. Quite a discrepancy? Yes, but his estimate was in fact based on scientific principles.

He used what is known as the "obliquity of the ecliptic" for his calculation. In our solar system, the Earth's orbital plane (the way it is going around the sun) is known as the ecliptic plane. Because the Earth is tilted on its axis, the way the Earth spins on a daily basis is not level with the ecliptic plane. The angle between the ecliptic plane and the Earth's rotational axis is called "axial tilt" or "the obliquity of the ecliptic."

The Earth currently has an axial tilt of about 23.5°. However, the obliquity of the ecliptic is not a fixed quantity but changes over time in a cycle with a period of 41,000 years. It is a very slow effect known as nutation, but over time, the changes can make a big difference in how the heavens are seen from Earth. A formula to determine how the heavens looked at given dates has been developed.[71]

The stone pillars that act as solar markers at the Kalasasaya temple at Tiwanaku, according to Posnansky's calculations using this formula, date the site, or at least the setting of these stones, at 15,000 BC. This proposal caused Posnansky's colleagues, especially in "western" countries, to balk; they slandered him, virtually destroying his career. They maintained that the Tiwanakans improperly placed the stones in their locations and that Posnansky was wrong.

Numerous elongated skulls have been found at Tiwanaku, and were displayed in the museum there until quite recently. No carbon 14 or DNA analysis has been conducted on them, so we have no idea how old they are.

As of October 2011 the display of skulls was made off limits to the public. Why? No reason was given, but this is the same thing we have seen (actually not seen!) at the museums in Lima; displays that once included elongated skulls have recently been removed.

Is there some sort of conspiracy at work here to keep the unusual elongated skulls from being viewed by the general public? Will they get the "wrong" ideas about the connection between ancient cultures that practiced cranial deformation? Or, even stranger, will people come to the conclusion, as have some scholars, that these skulls are of a different race than "normal" human beings?

The Incas and Cranial Deformation

And now we come to the Inca, who, as has been previously noted, most likely came from Tiwanaku, leaving some time around 900 AD, or perhaps as late as 1000. The most famous paintings of the Sapa (high ruling) Inca show them as having long dark hair. However, a close scrutiny of the early chronicles of the Spanish conquistadors reveal startling facts.

Francisco Pizarro, the leader of the Spanish who conquered the Inca beginning in 1532, had a secretary who, unlike Pizarro, could read and write. In his accounts, he states that at least some of the Inca had reddish to blondish hair, and it was cut short. One can still see today skulls in the basement museum of the Coricancha in Cusco, the central building of the Inca in their capital city, elongated skulls reputed to be those of Inca personages. If the stories are correct that the Inca were descended from the Tiwanaku, then they would have, to some degree, inherited physical traits—such as elongated skulls?

The civil war which occurred in Peru just prior to the arrival of the Spanish resulted in the death of most of the Inca at the hands of Atahualpa, the half-Inca who was on one side of this civil war. He was the younger son of the last of the true sovereign Sapa Inca, Huayna Capac, who, against previous cultural traditions, decided to divide the Inca civilization into two parts.

The Tahuantinsuyu as the Incan territory was called (meaning "four corners of the world") stretched from Colombia in the north to the center of what is now Chile to the south. In about 1527 Huayna Capac gave Colombia and Ecuador to his son Atahualpa, whose mother was of royal blood from the local area; the rest was the domain of Huascar, his firstborn son of pure Inca lineage. A truce lasted approximately five years and then Atahualpa decided that he wanted to take over the entire Tahuantinsuyu. He sent 40,000 troops south, and they captured Huascar in Cusco and imprisoned him. Then he called a meeting of all of the royal Inca family. Atahualpa decreed that now that Huascar had been removed from his place of power, a decision had to be made as to who was to succeed him. But this was a ruse on his part.[72]

As soon as all of the family had been assembled, the soldiers killed each and every one of them; men, women and children died. Thus, the vast majority of the royal Inca family, as much as 95 percent, were dead before the Spanish arrived. Oddly, no Inca mummies, at least ones that can clearly be identified as such, have ever been found. Oral traditions state that prior to the arrival of the Spanish all of the royal Inca mummies were removed from the capital Cuzco and hidden, lost to this day. Allegedly they were taken a secret city in the mountain jungles east of Cuzco called Paititi. Though searched for by explorers, Paititi has never been found.

Inca men with short hair are depicted in the drawings of Felipe Guaman Poma de Ayala, who chronicled the Inca after their fall, and published in his *Nueva Cronica Y Buen Gobierno* (*New Chronicles and Good Government*) in 1615. Is it possible that the short hair was a style meant to show off a cranial shape at least slightly different from the contemporary peoples'? What is curious is that the Paracas royalty, at least those examples of which are in the Paracas History Museum that still retain hair, were also closely cropped.

The ancient road system which is attributed to the Inca, stretching from Quito in the north to Santiago de Chile in the south, going along the Pacific coast, and through the highlands of the Andes, was not in fact wholly of Inca construction. The Wari, who occupied the Andes from Ayacucho (which is west of Cusco) south to Lake Titicaca, are known to have had an elaborate road system, which the Inca later adopted. These roads may easily predate the Wari as well.

So, we have seen that Tiwanaku may in fact be much older than conventional archaeology allows. There is mounting evidence that the oldest and most exacting stone constructions in Cusco predates the Inca by possibly thousands of years. The Paracas traded at least as far back as 1000 BC with people of the highlands and the Amazon jungle. A genetic relationship between the Paracas, Tiwanaku and Inca ancestors may be possible.

Could the cranial elongation phenomenon be one of many characteristics shared by these cultures over time and distance, and lead us to a common ancestral people? Could they be visitors from a distant nation, with highly evolved capabilities, reddish hair and artificially (or naturally) elongated skulls?

Such a hypothesis may be a stretch at present, and the lack of archaeological work in these areas stretch the concept even thinner, yet the anomalies compel us. Why elongated skulls? In order to resemble some ancestors that only exist in the mists of time? But ancestors with such highly evolved minds and sensibilities that their descendants, the Paracas, Tiwanakans and Inca, clearly left their marks on the landscape and in their graves—and captivate our interest in the 21st century? Only time, research and luck will allow us to answer these questions.

Diffusionism and South America

The origin of the Paracas people of the coast of Peru is completely unknown at this point. As has been pointed out, the only accredited scholar to study them in any way was Julio Tello in the 1920s, and his conclusions have simply been recycled in their entirety ever since. Tello's claims that ceramic styles and therefore cultural influence came to this part of Peru from the Chavin culture of the highlands has been claimed since his time as fact, when truthfully it can only be regarded as a supposition, and it still does not tell us where the original Paracas came from, and where their practice of cranial deformation originated.

It would seem that the Paracas people originated with ancient seafarers, rather then coming from the mountain regions. Tello and most early anthropologists were isolationists, rather than diffusionists. They saw ancient cultures in Peru as being isolated from the rest of the world, including Central America. The Olmecs as a culture, with their pervasive elongated crania, had not been "discovered" at the time of Tello's original work in Paracas, so he was unaware of the similarities between the Paracas culture and the Olmecs. Today, anthropologists are increasingly realizing that ancient cultures were far more connected than originally thought, and oceans are highways between civilizations rather than barriers. It would have been easy for the Olmecs to use boats to move south down the Pacific coast to South America They would never even have to leave the sight of land! It would almost seem impossible that they did not do this.

The idea that the Paracas people, both culturally and genetically, came from a source far from South America is tempting, though no specific source can be named at this time. Is it possible that they are related to the Olmecs, who lived to the north and were chronologically a contemporary culture? Possibly. The prevailing wind and ocean currents could and perhaps did take Paracas vessels as far north as Ecuador on trading missions, and it is not that much farther to the land of the Olmec. Once there, an approximate 100-mile trek takes you to the Caribbean shores of the Atlantic Ocean and yet more possibilities of diffusion of cultures. Perhaps it was the Olmecs who originally came south to the Pacific coasts of Ecuador, Peru and Chile on trading missions, bringing with them certain odd cultural customs, such as cranial deformation and pyramid building.

The Olmecs were also megalith builders as seen at the early Olmec city of Monte Alban and in the colossal basalt heads found near La Venta. Did the Olmecs also bring megalith building to South America?

Tiwanaku (Tiahuanaco) is believed by conventional scholars to have become a regional power about 300 AD, and thus it is presumed that the large megalithic buildings date from that time. However, archeologists are unable to explain where the technology came from to build the great stone structures, especially those at Puma Punku, a neighboring site consisting mainly of hard diorite "machined" stones that only tungsten carbide, carborundum or diamond tools could shape. The dating of the elongated skulls found there is unclear, and it seems that they have never been properly studied using C-14 or DNA analysis. No account has ever been published as far as we know.

And finally the Inca; even archaeological neophytes can tell that many of the megalithic structures in Cusco and the Sacred Valley (especially the Coricancha, Sachsayhuaman, Ollantaytambo and the unnamed shaped andesite-granite stone outcrops above Cusco) were clearly made by societies that had tools far harder and probably more complex than the bronze chisels and stone hammers of the Inca. These structures hint or even scream at us that another, older culture was involved. And what of the elongated skulls found in the basement museum of the Coricancha in Cusco, labelled "Inca" without any extensive explanation of who they were, how old they are or where they were found?

True Inca skeletons have never been found, having been hidden by the last of the Inca just before the Spanish ransacked Cusco in 1533. None of the Spanish chronicles, nor those of the Inca descendants, speak of them as being shorter than the Spanish, as are the vast majority of people inhabiting present-day Cusco. Hints suggest they stood out from the population of the time, perhaps having light coloured skin, and short reddish to even blond hair; it would seem they were descended at least in part from non-Native South Americans.

The so called Inca roads, such as those that connect Lake Titicaca (including Tiwanaku) with Cusco and the ancient city of Ayacucho, descending afterwards to the Pacific very close to Paracas, are far older than the Inca civilization. They date back at least as far as the Wari (Huari) culture, which held power from the 6th to 10th centuries AD, and may in fact be far older than that.

The Paracas presumably traded with people of the Amazon as far back as 3,000 years ago or more, as shown in the archaeological record by items such as weaving implements made of Chunta wood, which only exists in the dense rainforests east of the Andes. Also, parrot feathers and vicuna wool were used by the Paracas at that time in ceremonial mantles.

Is it possible that the elongated skull characteristic unites the Paracas, Tiwanakan and Inca bloodlines as coming from a common ancestor? Only more research will add fuel to this possibility. And as for connections with other cultures that displayed cranial deformation? The oceans of the world were never barriers to those of intelligence and at least an ounce of bravery. Let us treat these ancestral people with the respect they deserve, to have boldly lived, traded and loved to a greater extent than our narrow historical texts have thus far given them credit.

Top: The design often seen at Paracas, a man with two snakes coming out of his mouth and a UFO-bird on his forehead. A decapitated head as trophy is on his belt. *Bottom*: A trepanned skull discovered by Tello at Paracas.

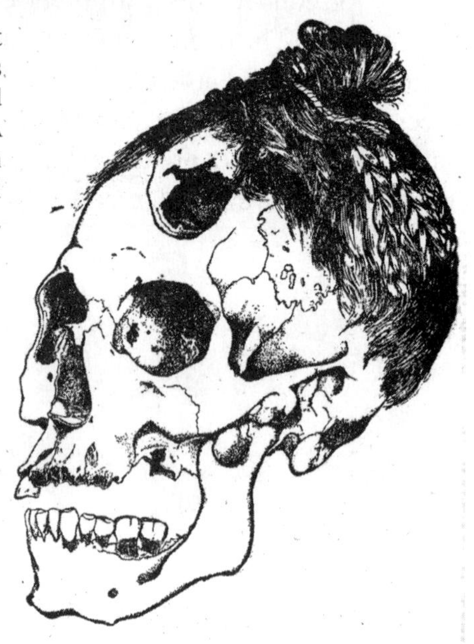

Two photos of the Chongos site at Paracas, as it looks today.

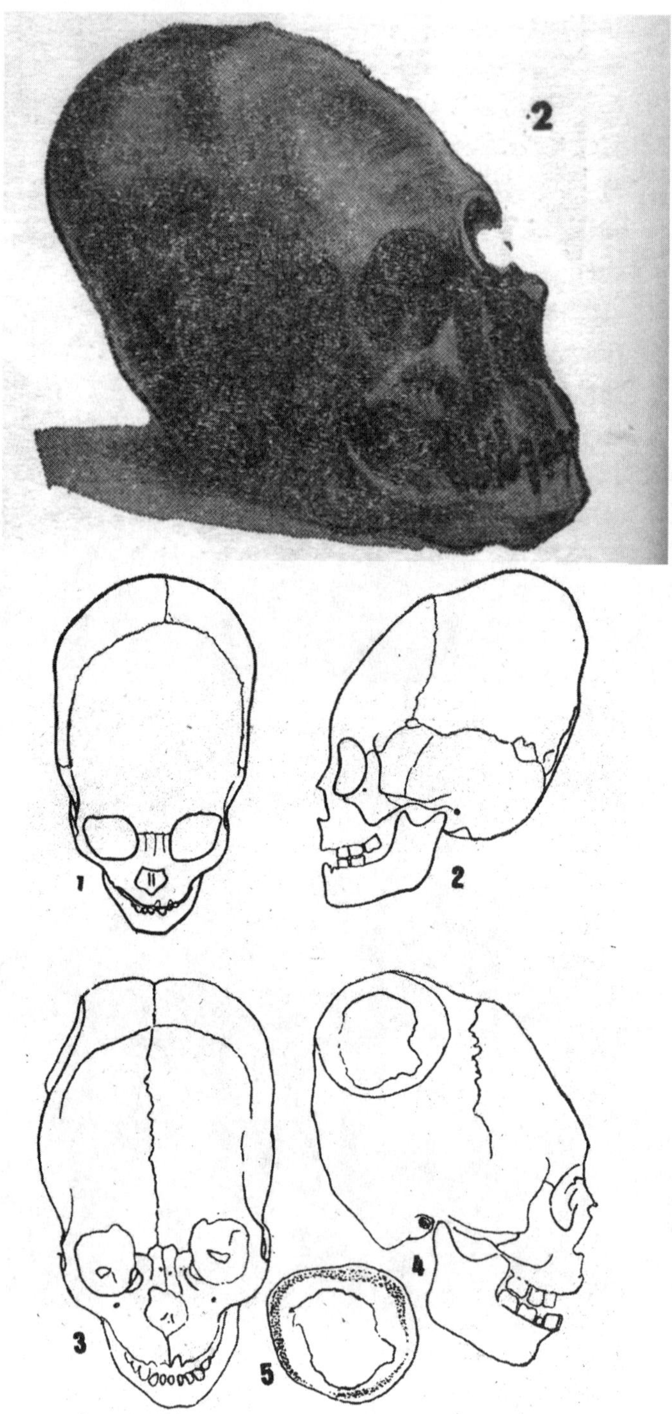

Early drawings of skulls from Cerro Colorado, Paracas by Julio Tello.

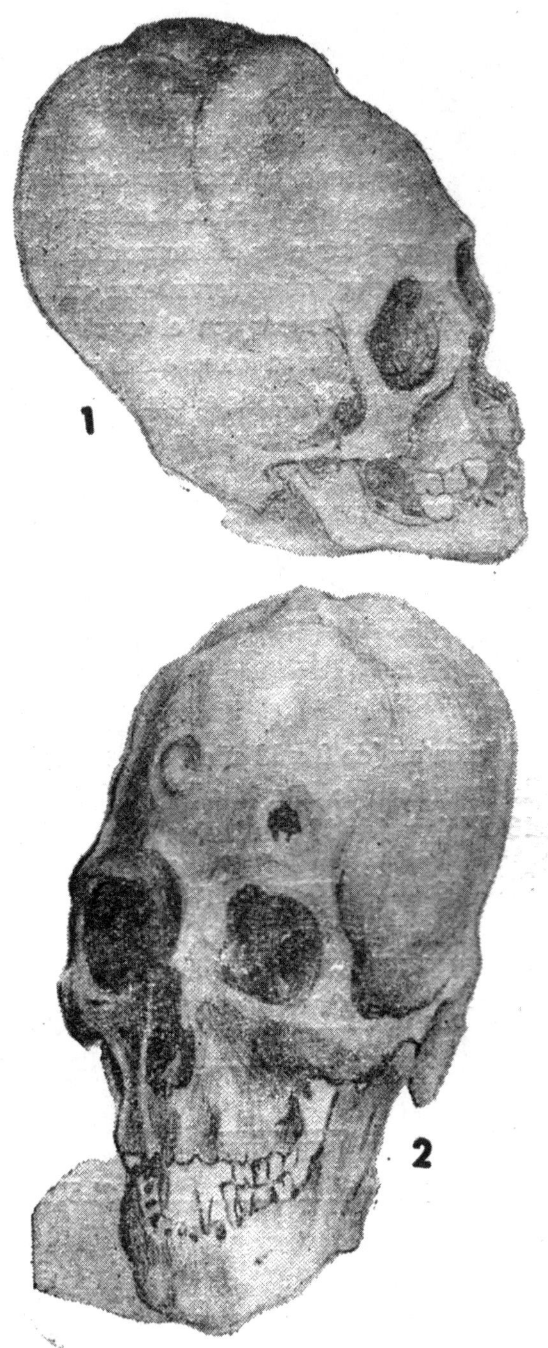

Early photos of elongated skulls from Cerro Colorado, Paracas by Julio Tello.

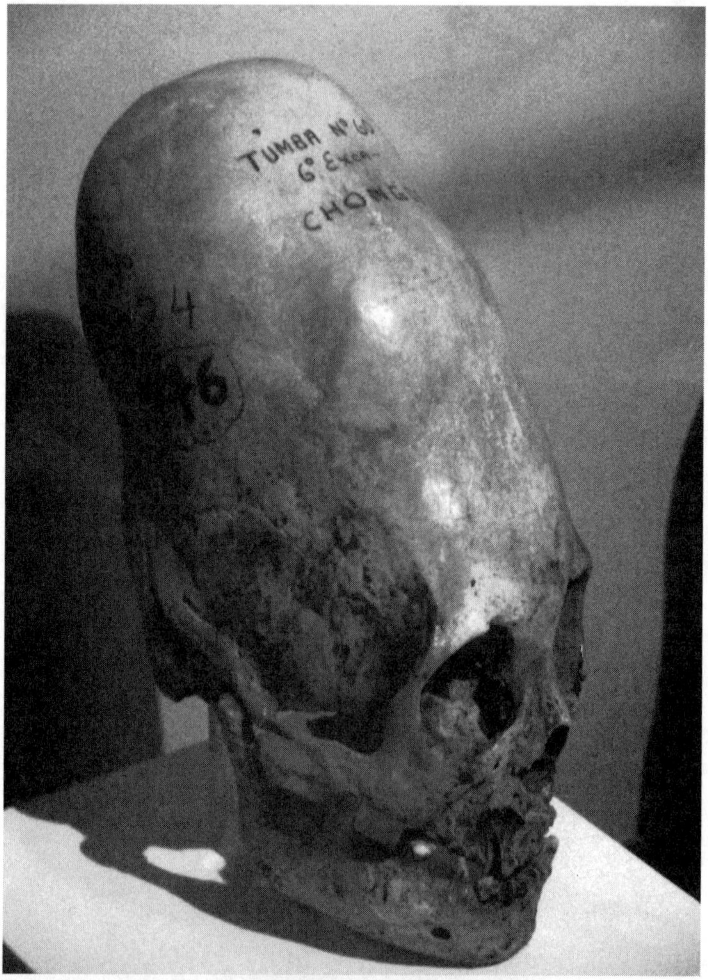

One of the extremely elongated skulls from the Chongos site near Paracas.

Paracas skulls on display at the Paracas History Museum.

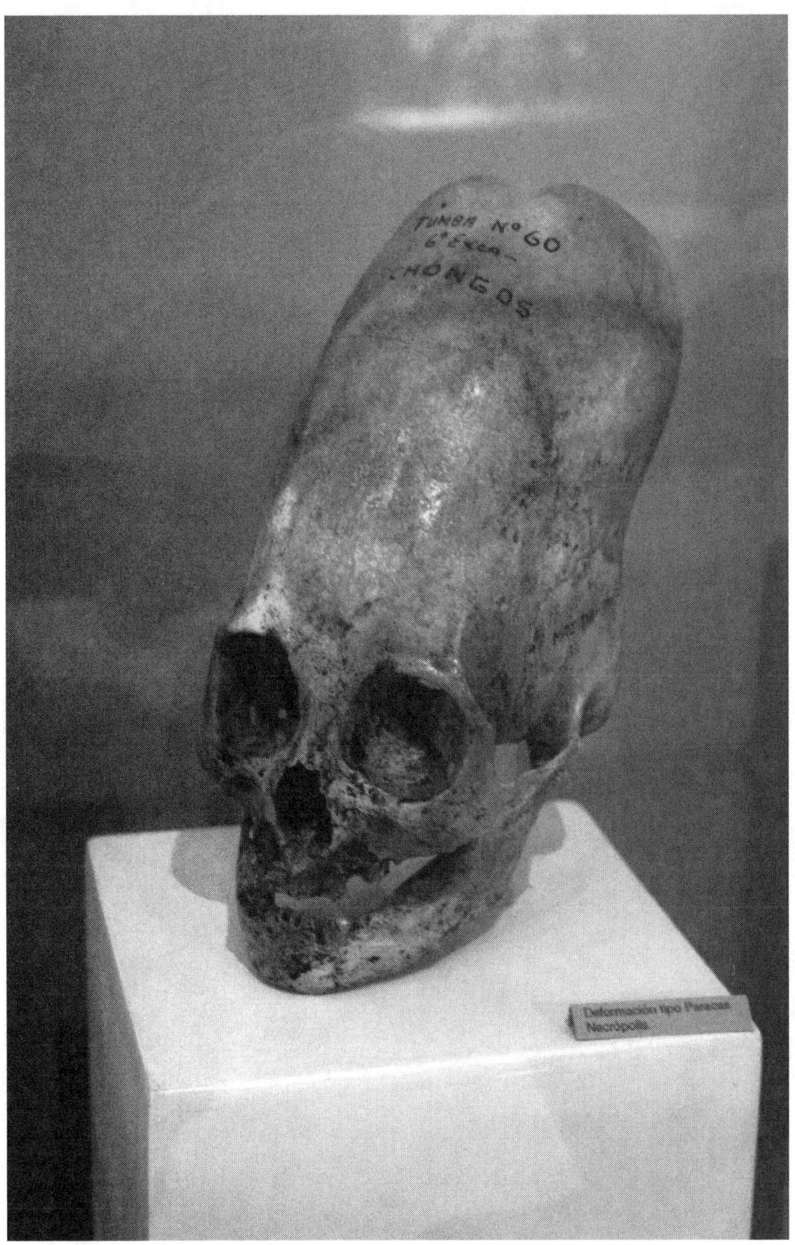

Another view of the extremely elongated skull from the Chongos site.

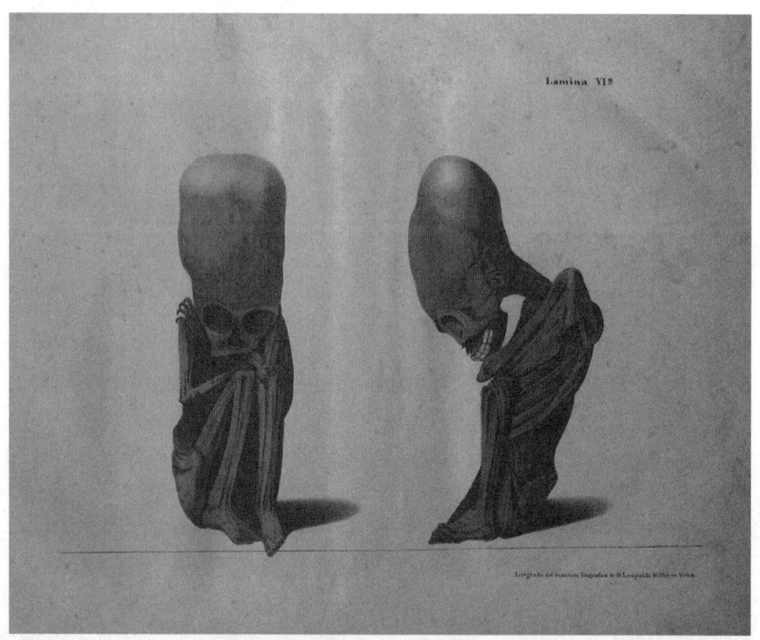

The illustration of elongated skull of a "fetus" from von Tschudi in 1851.

Many of the skulls from Paracas are kept in boxes in the back room as seen here.

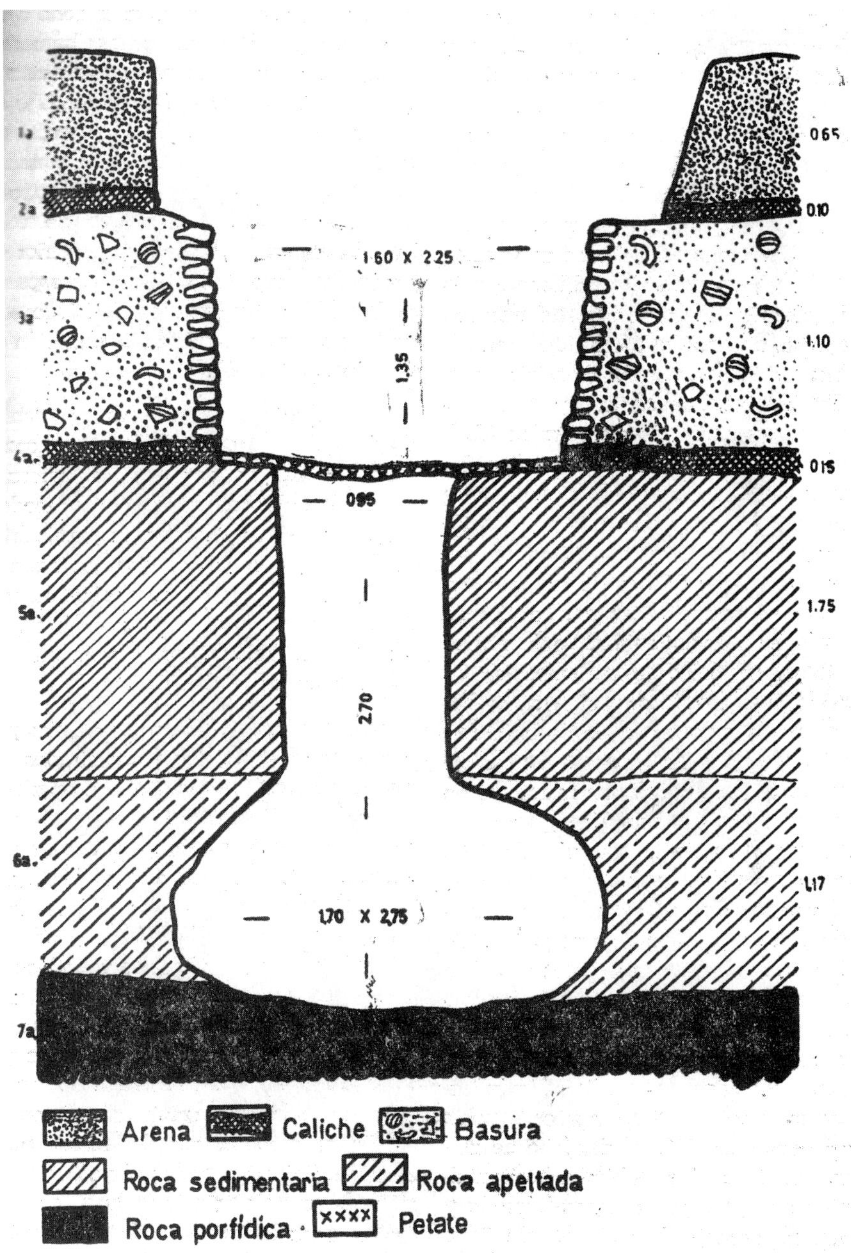

Julio Tello's drawing of one of the tombs he discovered at Cerro Colorado, Paracas.

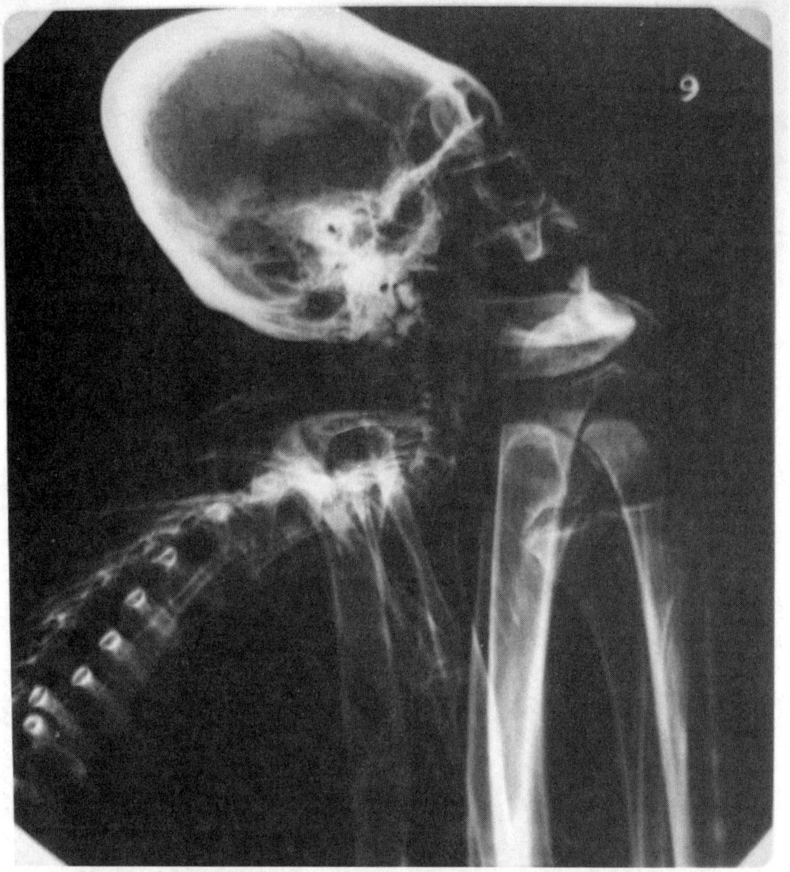

An X-ray of the one of the skeletons at the Archeology Museum in Ica.

An elongated skull can be seen on this "Inca" mummy from Nazca.

Massive elongated skull from Paracas now at the museum in Ica.

Another elongated skull from Paracas at the Ica museum, this one with hair on it.

More of the extremely elongated skulls from Paracas now at the Ica museum.

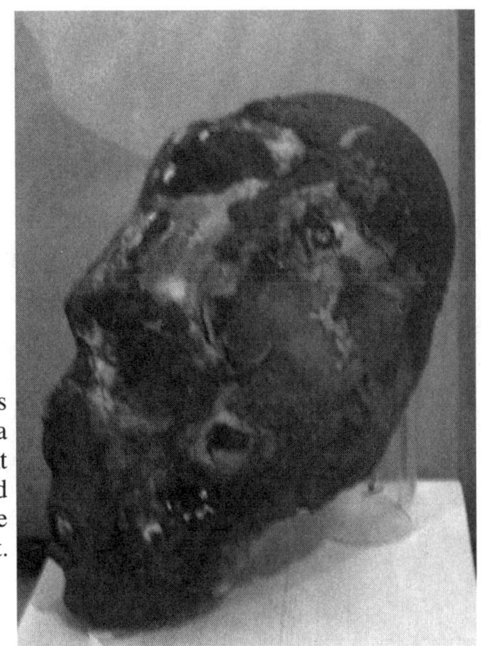

More of the elongated skulls from Paracas now at the Ica museum. The bottom right skull has been trepanned twice in the same area of the skull by the look of it.

The elongated Skull 13 at the Archeological Museum in Ica, near Paracas.

Elongated skulls once on display at the Tiwanaku Museum.

Elongated skulls once on display at the Tiwanaku Museum.

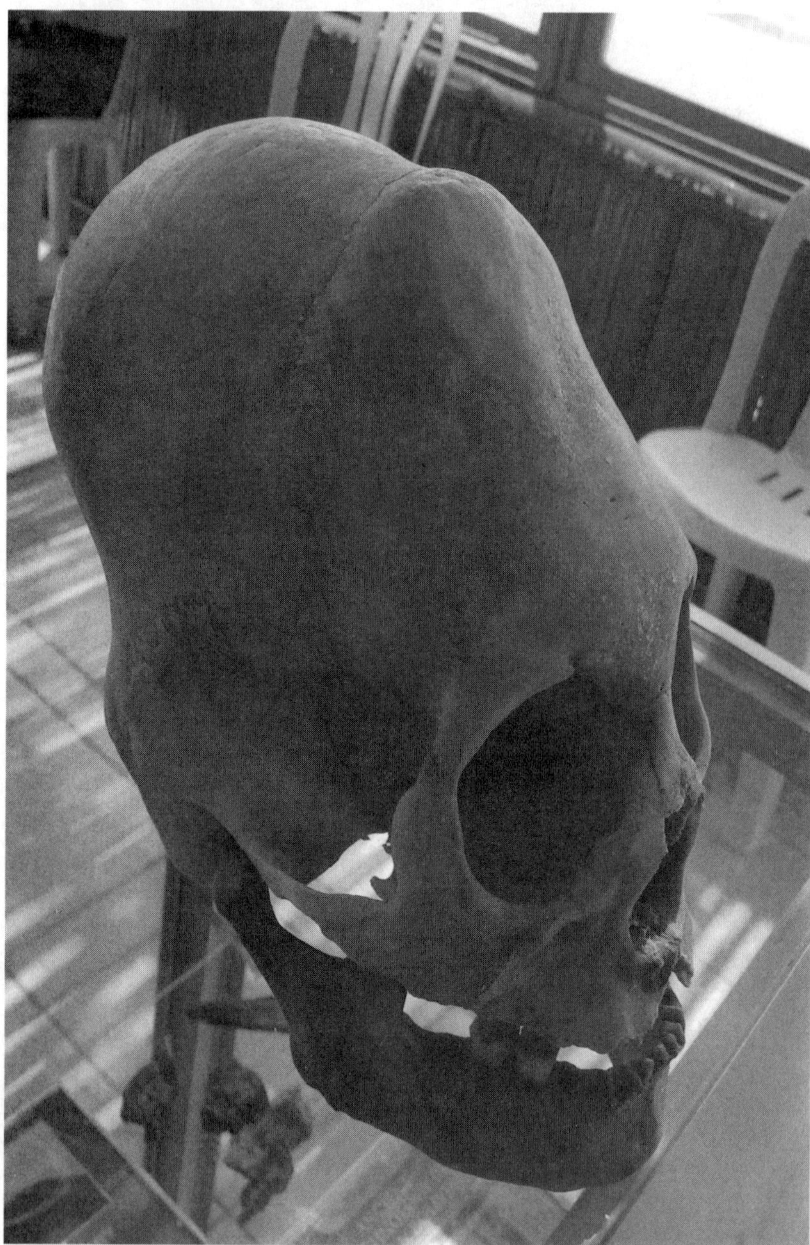

This page and opposite: One of the unusual skulls from Paracas. Elongated Skull form Camacho, near Paracas. The skull only has 2 major cranial plates, genetic orthodontic problems (missing molar teeth) and 2 curious holes in the back of the skull that may be foramen for blood and nerve flow.

Top of the skull showing only one parietal and one frontal plate.

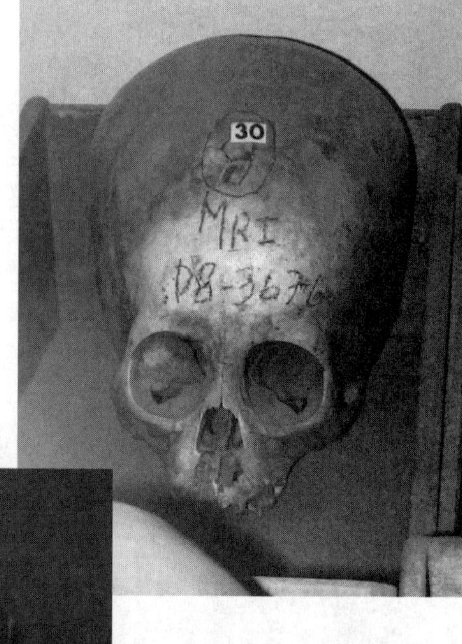

Above, left and below: Various elongated skulls now at the Paracas History Museum.

CHAPTER 5
EGYPTIAN, AFRICAN AND EUROPEAN CRANIAL DEFORMATION

If I could get my membership back,
I'd resign from the human race.
—Fred Allen

When we look at the Olmecs and the strange civilizations in South America, we have to wonder if there is not some connection to advanced ancient cultures in Africa, the Mediterranean and Asia. Was cranial deformation spread in ancient times by seafarers? It is known that cranial deformation was practiced in ancient Egypt and other areas of the Mediterranean. In fact, the earliest cranial deformation in the area may have been done on the island of Malta, near Sicily, but we will get to that later.

Cranial Deformation in Ancient Egypt

Cranial deformation was known in ancient Egypt, and it is difficult to say how many of the early kings and queens had elongated heads. We know that many of the Atonists (worshipers of the sun god Ra or Aton, followers of the Pharaoh Akhenaton) had elongated heads and are assumed to have attained them through headbinding. However, it is curious that we have no illustrations of headbinding in texts from ancient Egypt. This indicates that many things happened in ancient Egypt that we have no record of.

Akhenaton's wife, Queen Nefertiti is one of the most famous queens of ancient Egypt. Her beauty, revealed in her famous limestone portrait bust—one of the loveliest masterpieces of Egyptian sculpture—has made her widely known around the world. Yet, in spite of her fame, historians are uncertain about her origins. There appears to have been a deliberate attempt in ancient Egypt to erase the membory of her existence—in fact, the existence of all the Atonists. The powerful Amun priesthood had dominated Egypt before Akhenaton brought the Atonist movement forward, and quickly returned to power after his death. They appear to have tried to suppress any evidence of this break in the continuity of their control.

Nefertiti is a very mysterious figure. Historians disagree as to who she was, and who her parents were. Some historians believe she was just a commoner, while others have suggested that she was a Hittite princess, or similarly that she was a princess from Mitanni, a neighboring kingdom of the Hittites. Some Egyptologists believe that she was a daughter of Ay, one the viziers to the pharaoh. Clarifying the matter would help to clarify other significant aspects of ancient Egyptian civilization, one of which is why she seems to have had an elongated head!

An aspect of genetics, that appears not to have been given the attention it deserves, can help resolve this mystery. It is the elongated skull or the "dolichocephalic" head that many members of the eighteenth Egyptian dynasty possessed, including Nefertiti's famous husband Akhenaten, and her even more famous relative Tutankhamen. One of the reasons that historians ignored this deformation at first is because some thought that it was just a feature of stylized art. Some have suggested that elongated skulls are not an unusual feature, and prevail in some African and also Nordic tribes.

However, we are talking of a skull shape that goes well beyond the normal human shape, to the point that biologists have attributed it to a rare disease, some even to extraterrestrial sources.

Nefertiti, Akhenaton, and other Atonists can be seen in statues and wall paintings from Akhetaton (Amarna—the capital built by Akhenaton) to have clearly elongated skulls. Akhenaton and Nefertiti are thought to have had six children:

Meritaten, Meketaten, Ankhesenpaten (also known as Ankhesenamun, later queen of Tutankhamun), Neferneferuaten Tasherit, Neferneferure, and Setepenre. All of these children, according to the art of the time that has been preserved, had elongated crania. Tutankamun, whose original name was an Atonist name—Tutankhaten—also had an elongated skull, as statuary in his tomb demonstrates, as well as X-rays of his famous mummy. While not discussed much by Egyptologists, it is assumed that headbinding was used to elongate their skulls. Some authors have suggested, on the other hand, that Akhenaton was an extratrerrestrial and his skull was naturally elongated, rather than artificially.

If their skulls were deformed artificially, where did the Atonists get this custom? Was it brought from a foreign country? Some scholars think so, pointing to the Hittites and the land of Mitanni as the origin for this curious custom.

The Mysterious Asia Minor State of Mitanni

As we have said, some scholars believe that Nefertiti was not an Egyptian, but from Mitanni, an area of northern Iraq that was thought to be aligned with the Hittites and ancient India. As a Mitanni princess sent to Egypt for a diplomatic marriage and truce, Nefertiti may be one of the people who brought cranial deformation to ancient Egypt. At the time, Mitanni and the Hittites were in competition with Egypt for control of the territory which is now essentially the modern state of Syria.

So, in order to understand the Atonist custom of cranial deformation we need to have some understanding of the little-known ancient nation of Mitanni. Says the *Encyclopedia Britannica*:

> [Mitanni was an] Indo-Iranian empire centered in northern Mesopotamia that flourished from about 1500 to about 1360 BC. At its height the empire extended from Kirkuk (ancient Arrapkha) and the Zagros Mountains in the east through Assyria to the Mediterranean Sea in the west. Its heartland was the Khabur River region, where Wassukkani, its capital, was probably located.
>
> Mitanni was one of several kingdoms and small states (another being Hurri) founded by the Indo-Iranians in Mesopotamia and Syria. Although originally these Indo-Iranians were probably members of Aryan tribes that later settled in India, they apparently broke off from the main tribes on the way and migrated to Mesopotamia instead. There they settled among the Hurrian peoples and soon became the ruling noble class, called *maryannu*.
>
> The foreign policy of Mitanni during its early years was

> based largely on competition with Egypt for control of Syria, but amicable relations were established with the Egyptian king Thutmose IV (reigned 1425–17 BC). Perhaps the most outstanding Mitannian king was Saustatar (Shaushshatar; reigned *c.* 1500–*c.*1450 BC), who is said to have looted the Assyrian palace in Ashur. The last independent king of Mitanni was Tushratta (died *c.* 1360 BC), under whose reign Wassukkani was sacked by the Hittite king Suppiluliumas I. Tushratta was later assassinated, and dynastic struggles ensued until Mattiwaza, a son of Tushratta, was aided by Suppiluliumas against Shuttarna of Hurri; thereafter Mitanni became part of the Hittite empire and was called Hanigalbat. Shortly afterward, however, it was captured by the Assyrian Adad-nirari I (reigned *c.* 1307–*c.* 1275 BC) and again by Shalmaneser I (reigned *c.* 1275–*c.* 1245 BC), who turned the territory east of the Euphrates River into an Assyrian province.

From all the pieces of evidence that we can piece together, we can infer that the Indo-Iranians of Asia Minor were apparently in the habit of elongating skulls. Kurds from northern Iraq and eastern Turkey were said to practice cranial deformation until the 1940s. This custom apparently was practiced by the Kurds for thousands of years, going back to the empires of Mitanni and others circa 1500 BC. Cranial deformation was a common custom in this area among a number of different nations (going back to early Sumeria, circa 5000 BC) and Nefertiti and her entourage may have popularized the custom in Egypt.

Says the Hindu historian and engineer Dr. Ashok Malhotra at WorldHistoryBlog.com (after discussing cranial deformation) about Mitanni, Nefertiti and Queen Tiye (Akhenaton's mother):

> Nefertiti too possessed such a skull and therefore the possibility of her being a commoner becomes unlikely. The second speculation that she was a Hittite princess is also ruled out by reference to available historical records. Rather she appears to be a Mitanni princess daughter of the Mitanni king Dashrath. The confusion has arisen because

in historical records the Mitanni have been confused with Hittites on occasions. Both Hittites and the Mitanni belong to the Indo-European speaking Aryan races.

The Mitanni were a people of Aryan origin who ruled a vast kingdom with a largely Hurrian population in West Asia in the second millennium BC, for a brief historical epoch, sometime after 1500 BC. It was a feudal state led by a warrior nobility in which apparently the royal women were trained along with men in horse riding, chariot racing and warfare. This training was provided for the eventuality that they might be called upon to rule if widowed. Such accounts are found in the Puranas and Vedas, ancient historical records of the community that the Mitanni kings belonged too. The Rig-Veda, an ancient scripture of Mitanni rulers recounts the story of a warrior, Queen Vishpla, who lost her leg in battle, was fitted with iron prosthesis, and returned to battle. The Mitanni kingdom in Syria was a foreign and brief one lasting for about 150 years. During their brief reign the relationship they established with Egypt has left a significant mark in history. It was a mutually beneficial alliance that permitted the Mitanni to continue in foreign surroundings and provided a buffer to the Egyptians against Hittite incursions. The Mitanni kingdom was eventually weakened by Hittites and returned to Syria in approximately 1330 BC.

While they ruled in the area, the Mitanni Royal house developed close amicable relations with their western neighbors, the Egyptian Royal house through intermarriages as well as financial, military and religious alliances. For a period they became as one family. There appear to have been some alliances amongst the priestly class as well. The daughter of King Artatama was married to Thutmose IV, Akhenaten's grandfather. His son, King Shuttarna in the early fourteenth century BC sent his daughter Kiluhepa to Egypt for a marriage with Pharaoh Amenhotep III. And the daughter of the King Dasharatha, the son of Shuttarna, Princess Tadukhipa, became the queen of Akhenaten. [Note that the women's names were changed once they got to Egypt, which adds to any confusion.] The Egyptian Pharaohs also introduced horses and chariots in Egypt because of their relationship with the Mitannis.

The archeological finds at Amarna shed light on the relationship between the two royal families. In one Amarna letter, written to Akhenaten's mother, "Tiye, my sister," the Mitannian king complains that Akhenaten has not sent gifts that his father had promised, "I had asked your husband for statues of solid cast gold, but your son has sent me plated statues of wood. With gold being dirt in your son's country, why have they been a source of such distress to your son that he has not given them to me? Is this love?" Dashrath wrote to Tiye instead of to the pharaoh himself because he was more comfortable in writing to his sister than the king. The letter is hardly a diplomatic or royal letter. It is a family communication.

The origins of queen Tiye, like that of Nefertiti are also shrouded in controversy. It is very possible that the priests did not approve of the Egyptian family connection with Mitannis. They had good reasons for it. Primarily it was the introduction of foreign gods and unorthodox customs into Egypt as a result of these foreign queens. Queen Tiye too has been recognized for her unorthodoxy like Nefertiti. Historians have however admitted that there appears to be a relationship between Tiye and Nefertiti. There was. Tiye was Nefertiti's aunt – the sister of her father Dashrath. The Amarna letters prove the close family ties between Dashrath and Tiye. Another reason for the discomfort of the priesthood was that before the appearance of the Mitannis, the priestly clan often supplied brides to the pharaohs. That helped them to maintain their power in Egypt, but this new source of royal brides must have been a source of much anguish to the priestly clan. They may have responded by claiming that the new brides were not royal but just from a common tribal source that had managed to grab a neighboring kingdom. This last assumption may have arisen from their ignorance

> of Mitanni royal roots that have a history perhaps longer than even the Egyptian civilization as illustrated by their sacred texts, the Vedas.
>
> Some historians have claimed that Tiye was the daughter of Yuaa, a priest of Mitanni origin that her mother Tuaa, was of royal descent, from the royal family of Mitanni. If this latter was the case then it would make Tiye a cousin of King Dasharath rather than a blood sister. However, the utter informality of communications between Dasharath and Tiye, along with historical records indicating that the Mitanni kings had provided the Egyptian pharaohs with their daughters as queens suggests that Tiye was a blood sister of Dashrath, the Mitanni princess Kiluhepa. In either case the Mitanni royal origin of Tiye, and by extension that of Nefertiti appears to be of little doubt.

So, Dr. Malhotra, one of the few historians to discuss cranial deformation at any length, maintains that Nefertiti was indeed from Mitanni and that this Indo-Hindu culture was the origin of Egyptian skull deformation. Interestingly, Malhotra thinks that skull elongation is a genetic trait that would have been physically passed to the Egyptian royal house through their breeding with the Mitannis who possessed it. Atonism is a cosmology that probably originated in Asia Minor as well, and was brought to Egypt along with cranial deformation. The Hittites were known to be great seafarers, though many of their cities, including the famous Hattusas, were far inland. Still, they had many important ports such as Byblos along the eastern Mediterranean coast and controlled islands such as Cyprus.

Were the Hittites (and people from their satellite state, Mitanni) the ones who travelled originally on transoceanic voyages across the Atlantic and Pacific to bring the odd custom of cranial deformation to North and South America? We know that they had ships capable of making the voyage, and that Hittite culture was considerably advanced. They are often credited with "inventing" hard metals like iron and bronze. Were Atonist-Egyptians voyaging around the world spreading the custom of cranial deformation? The evidence would suggest that something exactly like this was occurring circa 1500 BC.

Certain ancient cultures were geographically located where they could have a port on two different oceans. We have already discussed how the Olmecs had important cities on both the Atlantic and Pacific coasts of Mexico. The Egyptians also had ports on the Red Sea (leading to the Indian Ocean) and the Mediterranean Sea (Atlantic Ocean) while the Hittites similarly had ports on both the Mediterranean and the Persian Gulf. This gives these civilizations the advantage of being able to trade by sea in two directions—east and west. Cranial deformation may have been disseminated around the world by these cultures. They may have penetrated deep into Africa as well. Let us now look at the curious case of the Mangbetu in central Africa.

Cranial Deformation in Central Africa

When early explorers and anthropologists penetrated much of the Congo River basin in the late 1800s they came across the Mangbetu tribe in what is today the northeast part of the Democratic Republic of the Congo. These early explorers were fascinated by the Mangbetu practice of head elongation. Babies' heads were bound with cloth to create the desired shape. As adults, the effect was emphasized by wrapping the hair around a woven basket frame so that the head appeared even more elongated.

According to Wikipedia the Mangbetu are:

> ...a people of the Democratic Republic of the Congo, living in Orientale Province. The majority live in the villages of Rungu, Poko, Watsa, Niangara, and Wamba.
>
> ...Mangbetu is a member of the Nilo-Saharan language group. The Mangbetu language is phonetically distinct from other languages in that it possesses both a voiced and a voiceless bilabial trill. The Labo language of Vanuatu, also known as Mewun, is the only other language known to possess this phonetic feature. The Mangbetu are known for their highly developed art and music. One instrument associated with and named after them is the Mangbetu harp or guitar. One harp has sold for over $100,000. Musicologists have also sought out the Mangbetu to make

> video and audio recordings of their music. The Mangbetu stood out to European explorers because of their elongated heads. Traditionally, babies' heads were wrapped tightly with cloth in order to give them this distinctive appearance. The practice began dying out in the 1950s with the arrival of more Europeans and westernization. Because of this distinctive look, it is easy to recognize Mangbetu figures in African art.

So, the Mangbetu are a people who known for their elongated heads. They are also known, curiously, for their high knowledge of music and musical instruments. Were they taught by some elite group who also taught them the practice of headbinding?

Their language is likened to the Labo language of Vanuatu; what is intriguing is that people of Vanuatu also practiced head elongation, which we shall see later.

The Wikipedia post includes a quote from Prof. Peter J. Bloom's book entitled *French Colonial Documentary: Mythologies of Humanitarianism:* "However, it is part of an open text in which the lineage of the Mangbetu has been traced to Egypt, particularly to the ancient Egyptian iconography of Nefertiti."

Traditionally, the Mangbetu are said to have migrated from Sudan around the middle of the 18th century, coming into contact with other tribes on their way to their current homeland, where they settled in the 19th century. They appear to have imposed their language and customs on the surrounding tribes, the Mundu, Abisanga, etc. The name Mangbetu applies only to the ruling aristocracy and it is interesting to note the similarities between this situation and that of the Mitanni, who halted their migration in the Fertile Crescent and took control of the local cultures there. Once a considerable power, they have practically disappeared; their language and culture, however, remain, maintained by their subjects, with whom they have to a large extent intermixed.

The above is very intriguing because, if the royal Mangbetu truly originated from Sudan, which is just south of Egypt, they could very possibly be descendants of the ancient Egyptians, as in the Amarna period. Did some of the Atonists migrate into Sudan and eventually to the northeast edge of the Congo? Or, had ancient seafarers like the Hittites penetrated deep into Central Africa via the Congo river system? Physically the Mangbetu differ greatly from their neighbors. Their skin is not as black and their faces are less Negroid in appearance with many having quite aquiline noses. The spread of cranial deformation is fascinating and confusing as we have seen.

The Serpent Priests of Malta

Perhaps related to the ancient Hittites is the island of Malta, known for it extremely old megalithic temples and underground rock-cut temples. Early archeologists made a number of unusual skeletal discoveries during excavations after World War I, including skulls with elongated crania.

In a fascinating article by Italian archeologist Adriano Forgione published in *Hera* magazine (1999, Rome, Italy) it is mentioned that a number of dolichocephalous skulls were discovered in a very ancient underground temple on the Mediterranean island of Malta called the hypogeum.

According to Forgione:

> It was known that until 1985 a number of skulls, found in pre-historic Maltese temples at Taxien, Ggantja and Hal Saflieni, were exposed [displayed] in the Archeological Museum of the Valletta. But since a few years ago, these were removed and placed in the deposits. From then, they were not to be seen by the public. Only the photographs taken by the Maltese researcher Dr. Anton Mifsud and his colleague, Dr. Charles Savona Ventura, remained to testify the existence of the skulls and prove their abnormality. Books written by the two Maltese doctors, who since our first day in Malta helpfully provided us the necessary documentation for our research, illustrate a collection of skulls that show peculiar abnormalities and/or pathologies. Sometimes inexistent cranial knitting lines [sutures], abnormally developed temporal partitions, drilled and swollen occupants as following recovered traumas, but above all, a strange, lengthened skull, bigger and more peculiar than the others, lacking of the median knitting. The presence of this finding leads to a number of possible hypotheses. The

similitude with other similar skulls, from Egypt to South America, the particular deformity, unique in the panorama of medical pathology referred to such distant times (we are talking about approximately 3000 years before Christ) could be an exceptional discovery. Was that skull a result of ancient genetic mutation between different races that lived on that island?

Located right in the middle of the Mediterranean, the islands of Malta and Gozo have been strategically important throughout history. They have been controlled by every major civilization to gain hegemony in the region, and played important roles in the Crusades and WWII. Forgione says that the islands were also very important centers in prehistoric times, places where "medical cures" were conducted, oracles were consulted and ritual encounters with the priests of the goddess took place. There existed many sanctuaries and thaumaturgic centers, where priests surrounded the healing goddess. He points out that it is well known that, in antiquity, the serpent was associated with the goddess and with healing capacities. The snake also belongs to the subterranean world. Therefore, a hypogeum dedicated to the goddess and the water cult was the perfect place for a sacerdotal group that was defined, in all the most ancient cultures, as the "serpent priests."

Says Forgione:

> Perhaps the skulls found in the hypogeum and examined during our visit to Malta, belonged indeed to these priests. As mentioned before, they present an accentuated dolichocephalous, which is particularly the center of our analysis. The long head and drawn features must have given a serpent-like appearance, stretching the eyes and skin. Lacking the lower part of the [skeleton], we can only speculate, but the hypothesis can't be far from reality, a reality worsened by the fact that such deformities certainly created walking problems, forcing him to slither! The lack of the cranium's median knitting and therefore, the impossibility of the brain's consistent, radial expansion in the skullcap, [meant] that it developed in the occipital zone of the cerebellum, deforming

> the cranium that looked like a single cap from the frontal and occipital area. This must have certainly caused the man terrible agony since infancy, but probably enhanced visions that were considered as being proof of a bond with the goddess.

Okay, maybe Forgione overdramatizes in making his case that the Maltese coneheads were literally "serpent" priests. As we have noted, the practice of cranial deformation has continued into the modern day, and there have been no anthropological reports that people with elongated heads were forced to "slither," or that their brains could not develop and they experienced excruciating, constant pain. Furthermore, Forgione believes that the elongation of the Maltese skulls is natural; it hardly seems likely that a widespread natural trait would cause the people who exhibit it such distress. But his point is well taken that the elongation of the head would set apart a certain class, and their distinctive look could tend toward the serpentine.

Skulls From a Different Race?

Incredibly, it may be that the elongated skulls from Malta are from a race other than *Homo sapiens*. Forgione says, "Even the other skulls we examined presented strange anomalies. Some were more natural and harmonic than the cranium that mostly gained our attention, but they still presented a pronounced natural dolichocephalous and we could assume, without fear of refutation, that it is distinctive of an actual race, different to the native populations of Malta and Gozo."

This possibility was confirmed by the Maltese archeologists themselves, Anthony Buonanno and Mark Anthony Mifsud, who told Forgione: "They are another race although C-14 or DNA exams haven't yet been performed. Perhaps these individuals originated from Sicily."

Continues Forgione:

> A fair part of the 7000 skeletons dug out of the Hal Saflieni hypogeum and examined by Themistocles Zammit in 1921, present artificially performed deformations. A skeleton of the group that was unburied by the archeologist, Brochtorff Circle, shows clear signs of intentional deformation through bondage. These deformations occurred for various reasons: initiations,

> matrimonies, solar rituals or punishments for social crimes or transgressions. All the tribal apparatus of incisions, perforations, partial or total removals, cauterizations, abrasions, insertions of extraneous bodies in muscles, like the modification of bodies for magical, medical or cosmetic purposes, were part of cruel practices in such, but 'with best intention' for the community. Why such persistence in tormenting one's own body? Was there any connection between the tribal rituals and the men of the lengthened cranium? Could it be possible that, as in other cultures, successive populations tended to deform their infants' heads in order to make them similar to this race of 'serpent priests'? In Malta, all this was practiced by a mysterious populace that erected gigantic temples to the Mother Goddess between 4100 and 2500 B.C. The presence of these skulls might be that of the last exponents of the most ancient sacerdotal caste that built the megalithic temples and, never having blended with the local populations, had continued reproducing through the millenniums [sic] within familiar unions (as was the usual practice among the elite) and consequently impoverished its genetic patrimony until inevitable pathologies manifested, finally disappearing.

So, in Forgione's estimation, a different race of people who had elongated heads came to Malta and built the ancient megalithic temples there between 4100 and 2500 BC. They were considered priests by the local population, with whom they did not interbreed. Their continued inbreeding inevitably resulted in the weakening of their gene pool and their demise, but they were so revered by the regular people that their strange characteristics were imitated by means of ritual deformation.

In fact, the ruins at Malta are now thought, even by mainstream archeologists, to date back to 7000 BC or earlier. Since Malta and the surrounding islands have been continuously occupied, the skulls could date from almost any time period, but one would suspect they are very ancient, especially given the locations at which they were found. At possibly 9,000 years old, these skulls may be of some different human species—or possibly of some extraterrestrial or hybrid race—that was interacting with *Homo sapiens* and other hominids, even though Forgione thinks they did not intermingle.

Cranial deformation is little studied and we are just beginning to explore the enigma of elongated heads. It remains to be seen if it can be proven that the Maltese skulls come from a different race. They are downright weird, whatever they are!

In the future, much-needed DNA testing on elongated skulls can resolve many of the questions. If the DNA on some of the stranger specimens, such as those found at Paracas in Peru, can be proven to be different from that of a normal human, the academic community and the media will be forced to take a deeper interest in the subject.

The Huns and Headbinding

Although it has not been established with certainty, some of the Huns are thought to be descendants of the Xiongnu (or Hsiung-nu) rulers, whose people lived to the north of China in what is now Mongolia 300 years before the emergence of the Huns. When that state collapsed in the first century AD, the Huns are thought to have migrated to Central Asia, intermingling with various Siberian, Turkic, Ugric and Iranian ethnicities. It is from Central Asia that they launched their fierce drive into Europe that helped bring down the Roman Empire. Says the *Brittanica Concise Encyclopedia*:

> Appearing from central Asia after the mid-4th century, they first overran the Alani, who occupied the plains between the Volga and Don rivers, and then overthrew the Ostrogoths living between the Don and Dniester rivers. About 376 they defeated the Visigoths living in what is now approximately Romania and reached the Danubian frontier of the Roman Empire. As warriors, they inspired almost unparalleled fear throughout Europe; they were accurate mounted archers, and their rapid, ferocious charges brought them overwhelming victories They extended their power over many of the Germanic peoples of central Europe and allied themselves with the Romans. By 432 the leadership of various groups of Huns had been centralized under a single king, Rua (Rugila). After his death (434), he was succeeded by his two nephews, Bleda and Attila.

Just about everyone has heard of Attila the Hun. His name is still synonymous with brutal conquest. Says the *Oxford Companion to*

Classical Literature:

> Huns [were] nomadic people, formidable horsemen, originating in central Asia, described by the contemporary historian Ammianus Marcellinus as savages who drank the blood of their slaughtered enemies. Their advance westward in the fourth century AD to the area north of the Black Sea caused terror and the widespread migration of barbarian peoples across the frontiers of the Roman empire. ...

The "widespread migration of barbarian peoples" referenced above is known as the Great Migration, and it is generally agreed that it was the Huns who drove these people from the periphery of the Roman Empire over its borders into previously controlled areas. These people were not easily controlled (being barbarians, after all) and unrest and uprisings eventually caused the disintegration of Roman rule.

So what caliber of people is it that can cause barbarian tribes to flee before them? The *Gale Encyclopedia of Occultism and Paraspychology* has this to say:

> Ancient historians recorded legends that grew out of the severe stress the Huns created in all those whom they fought against. They credited the Huns with a supernatural origin. The Huns were referred to as "children of the devil," because it was said that they were born of a union between demons and hideous witches, the latter cast out of their own country by Philimer, king of the Goths, and his army. The old writers state that the Huns were of horrible deformity and could not be mistaken for anything but the children of demons.

Two of the ancient historians referenced above are Jordanes and Ammianus Marcellinus. While Jordanes wrote his history of Rome around 550 AD, Ammianus Marcellinus was an actual contemporary of the Huns, writing in the latter part of the fourth century AD. He did not have much nice to say, which was actually pretty standard for Roman historians discussing enemies. According to Wikipedia:

> Jordanes and Ammianus report that the Huns practiced

> scarification, slashing the faces of the male infants with swords to discourage beard growth. Another custom of the Huns was to strap their children's noses flat from an early age, in order to widen their faces, as to increase the terror their looks instilled upon their enemies. Certain Hun skeletons have shown evidence of artificially deformed skulls that are a result of ritual head binding at a young age.

The Wikipedia article cites Peter Delius and the National Geographic book *Visual History of the World* with the comment about the headbound skeletons. In the case of the Huns, there is plentiful funerary evidence of cranial deformation that has survived to be excavated. There is an interesting article on the website forums.skadl.net entitled "On Cranial Artificial Deformation in Germanic Medieval Populations." Under the subtitle "Hunnic Cranial Deformation," the article states that one to five percent of burials in western European cemeteries of Germanic groups of the fourth to sixth centuries AD, many of which were influenced by the Huns as we have seen, exhibit cranial deformations. However, in Alani and Sarmatian burials in eastern European cemeteries of the second to fourth centuries AD, 80% of skeletons exhibit deformed crania. You may recall that the Alani were the first group overtaken by the Huns.

Did the Huns get the idea of headbinding from the Alani, and then pass it to other Germanic tribes who continued the practice for a few centuries? It is interesting that the Huns appear to have adopted bodily deformation to instill fear in the hearts of their enemies and appear more formidable in battle—not a bad stratagem for a marauding band! It obviously worked, as we invoke their name to this day to describe people who barbarically wreak havoc wherever they go.

Other Elongated Skulls in Europe

It is thought that cranial deformation was practiced in ancient times in many parts of Europe. Elongated skulls are thought to have been occasionally discovered by early archeologists of the Middle Ages, and reported by them to be the skulls of "giants." When an unusually large skull is discovered (and elongated skulls are typically quite big) it is often thought to be from a giant. Indeed, many of these people were unusually tall to begin with, and an elongated skull would add to a person's height.

In some parts of Europe, especially France, head elongation was practiced up until the late 19th century. In the Deux-Sevres area, head elongation involved wrapping the baby's head in a tight bandage. The binding was left in place for a period of two to four months and was then replaced with a fitted basket. When the baby was older, the basket was strengthened with metal thread. Hungarians were also known to use headbinding as a way of elongating the head.

As it turns out, headbinding gained popularity with the advancement in the mid-1800s of the theory of phrenology. In phrenology, the shape of the head was the determinant of a person's personality attributes; there were criminal shapes, intellectual shapes and so on. Belief in this pseudo-science led to a fad of headbinding infants in order to produce the proper shape for intelligent, well-adjusted growth.

A coneheaded skull is on display in the museum of Tuchersfeld, Germany. The inscription states that about 20 similar male and female skulls were found in the Franconia-Suisse area of Bavaria. This is one of the only elongated skulls available for viewing at any museum in Europe.

The Russian news outlet *Pravda News* reported on October 6, 2005 that extremely dolichocephalic heads had been discovered in the Russian Caucasus. The "Pyatigorsk" skull was found near Kislovodsk and dates from between the 3rd and 5th century AD. Said Pravda:

> "Researchers have repeatedly proved that the skulls had been deformed on purpose," said Dr. Kuznestov. "Ropes or special blocks were tied tightly round the heads of infants, over the temples. The custom went out of fashion by the 17th century. The reason behind the deformation phenomenon is still unknown. It is hard to say whether the methods worked effectively or not, since nobody ever conducted scientific experiments regarding the binding of the infants' heads."

Regardless of whether this skull was deformed by headbinding, it bore a striking resemblance to the skulls of Peru.

In January 2009 it was reported on digitaljournal.com that extreme dolichocephalic skulls had been dug up in Omsk, Siberia by Russian archaeologists. A one-minute video purportedly shows skulls dating from the 4th century AD that were being studied by scholars at the Omsk Museum of History and Culture.

Igor Skandakov, director of the museum, said that the skulls have marks that could be evidence of artificial deformation of a normal skull. He claimed that the skull was kept away from the public because of its unusual shape that shocked and frightened people.

Archaeologist Alexi Matveyev hypothesized that the deformation was carried out as a status symbol of belonging to the elite of society, or as a way to enhance brain function. "It's unlikely that the ancients knew much about neuro-surgery. But it's possible that somehow they were able to develop exceptional brain capabilities."

So we see that the phenomenon existed in Europe and receives the same type of treatment as in other cultures: vague speculation and a lack of extensive research, and a certain tendency to try to sweep it in the back room. Let us now look to the Far East and Pacific in our quest to uncover the enigma of cranial deformation and coneheads.

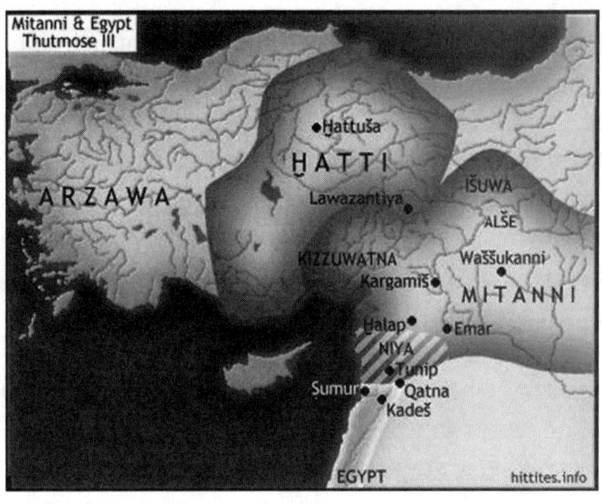

Two of the daughters of Akhenaton and Nefertiti, Meketaton and Meritaton, with their elongated cranium.

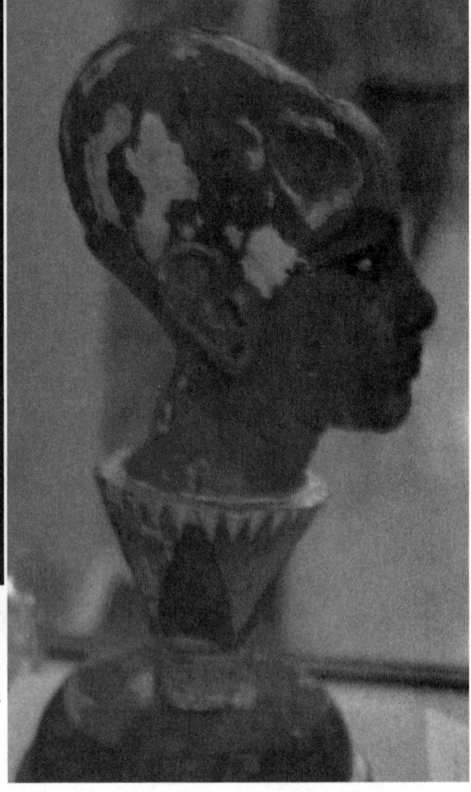

Top: A baslt statuette of Meretaton, one of the daughters of Akhenaton and Nefertiti, showing her highly elongated skull. *Above and right*: A bust of Tutankhamum showing his elongated head.

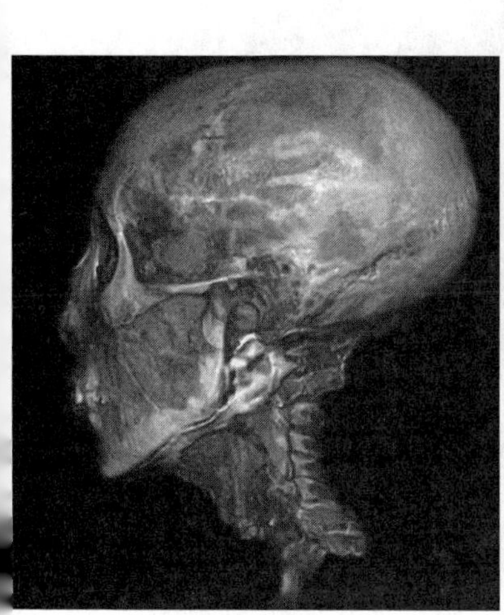

An X-ray of Tutankhamun's head.

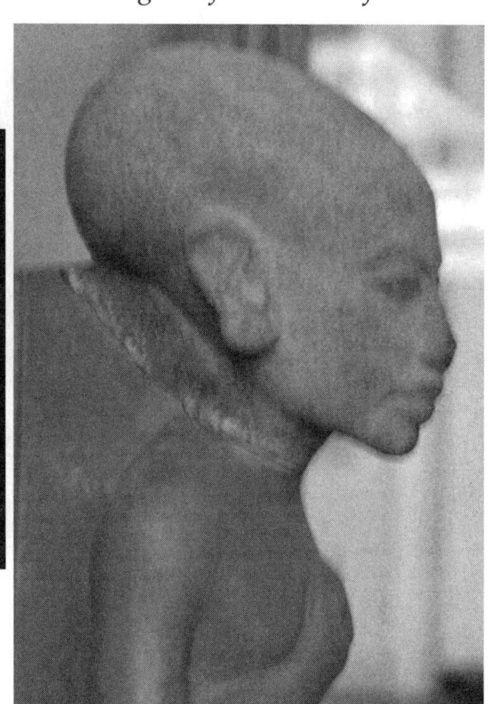

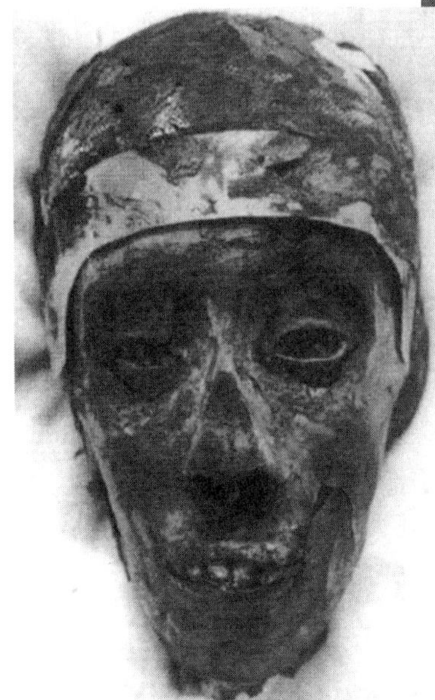

Top right: A statuette of Tutankhamum showing how he looked in life with a highly elongated head. *Below*: Two photos of the mummified head of Tutankhamum. From these angles, the cranial deformation is difficult to see.

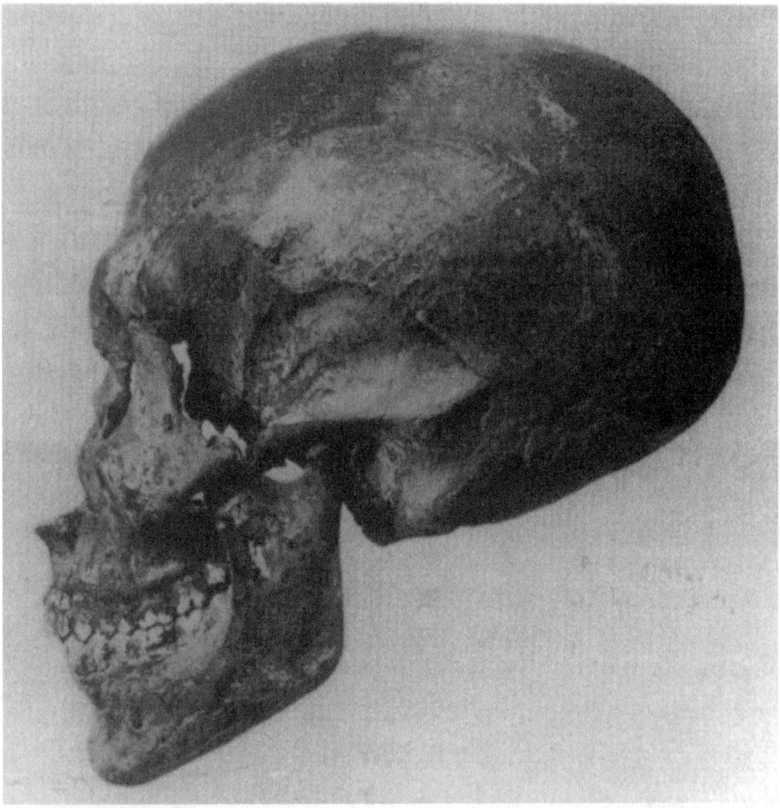

The elongated cranium of Smenkhare, brother (?) of Tutankhamun.

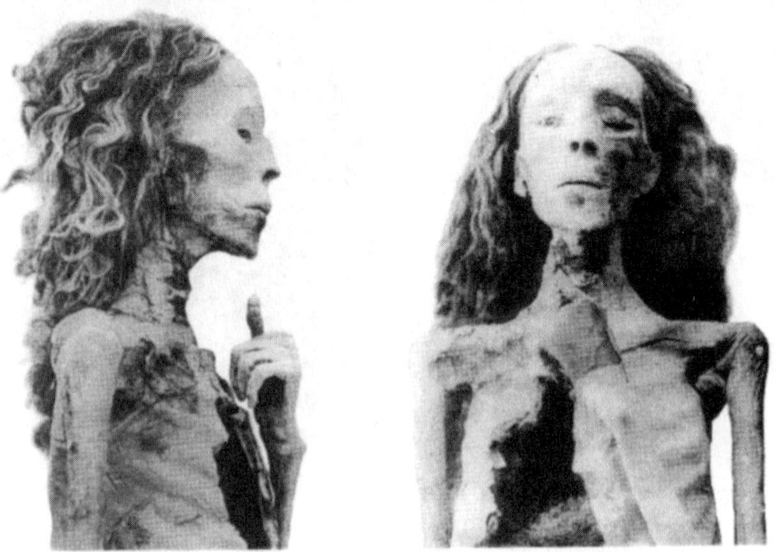

The mummy of Queen Tiye; her hair obscures what looks like an elongated skull.

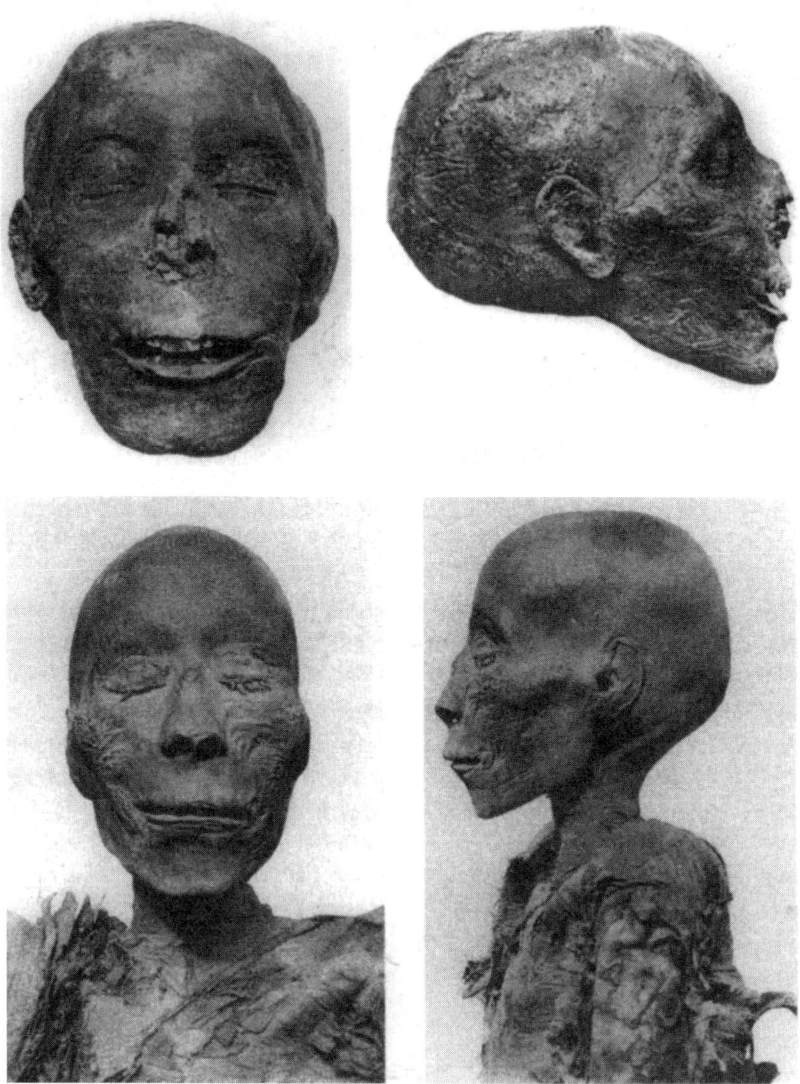

Top: The mummifed head of Thutmose I. *Below*: The mummified head of Thutmose III. Both seem to have deformed crania, particularly Thutmose I.

Top: The second coffin of Yuya, outside his tomb in the Valley of Kings. *Below*: The mummified head of Yuya. He appears to have a slightly elongated cranium.

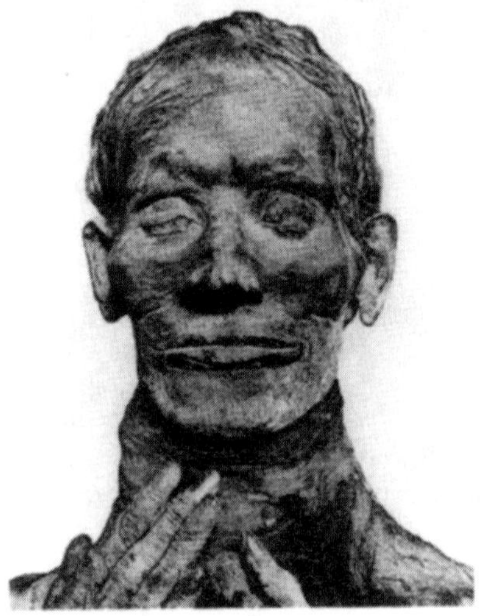

A clay cuneiform letter from Mitanni to Amenhotep III, one of the Amarna Letters.

A trepanned skull with the skin having grown over the hole, at the Archeology Museum in Aswan, Egypt.

Mangbetu child with an elongated skull and musical instrument.

Mangbetu mother with an enfant whose head has been bound, circa 1930.

Mangbetu woman with the elongated skull, circa 1930.

Mangbetu chief with a special status knife.

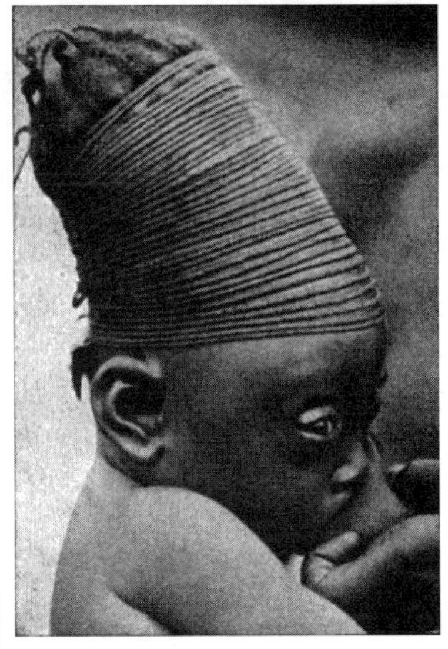

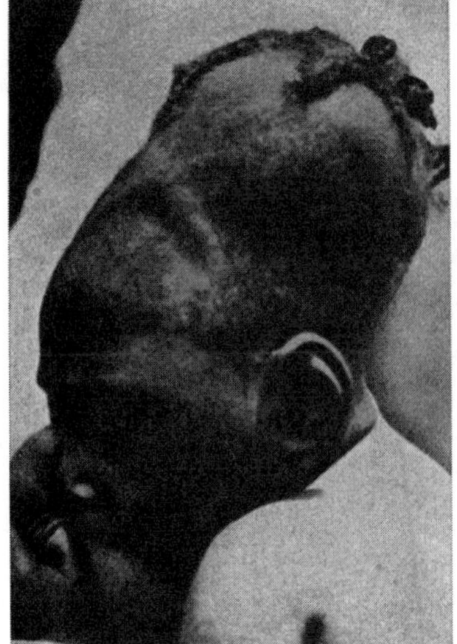

Above and left: Mangbetu babies, circa 1930s. One has had its headbing removed.

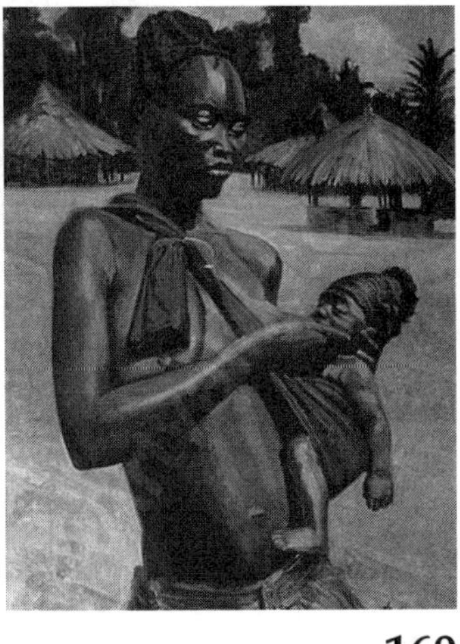

Mangbetu mother with a headbound baby in a sling

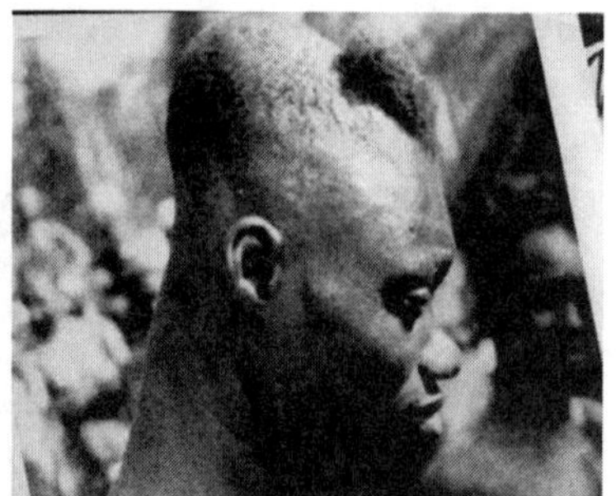

Mangbetu man with an elongated head.

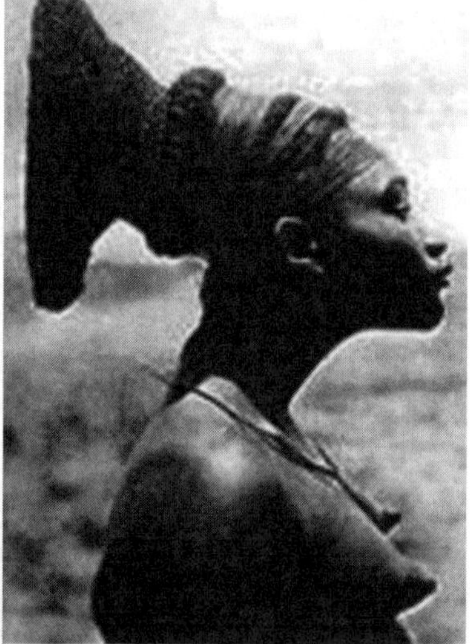

Mangbetu woman with an elongated head wearing a hairstyle and basketry that accentuate the effect.

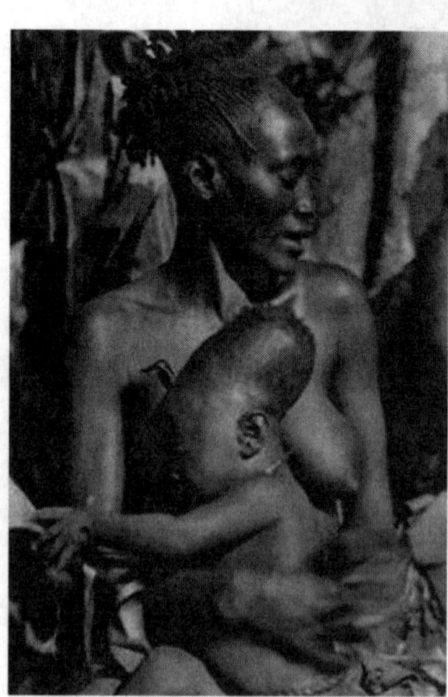

Mangbetu mother with an infant whose skull has been elongated.

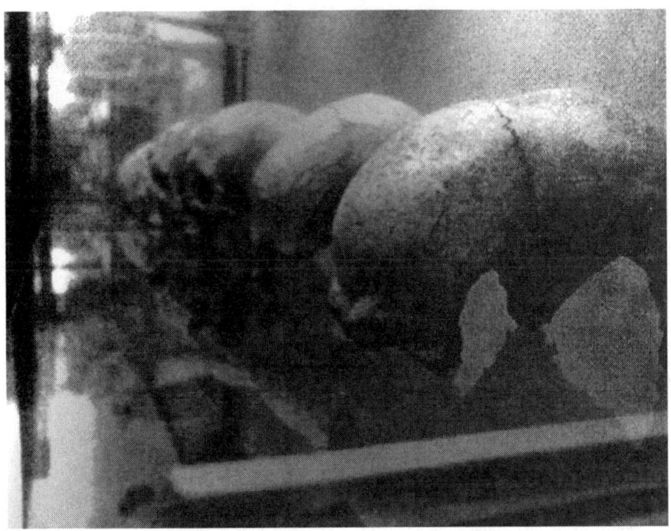

Rare photo of the elongated Maltese skulls once on display in Valetta. Thanks to Austrian researcher Klaus Dona.

THE MYSTERY OF MALTA'S LONG-HEADED SKULLS

The Editor of HERA, Italy's Magazine of Ancient Mysteries, Unravels One of the World's Strangest Puzzles

Long ago in the megalithic temple of Hal Saflieni, in Malta, were buried men with extraordinary cranial volume. Their skulls seem to observers today to belong to a truly alien stock. If properly understood, could these skulls provide a link between the ancient Mesopotamian and Egyptian cultures and a race of sacerdotal (priestly) men identified with the serpent?

We are back from a visit in Malta where we traced an ancient tale originating from megalithic temples to the Mother Goddess—adored in prehistoric times when the island was an important cult center. We were trying to resolve a mystery of truly stunning implications.

Before 1985 a number of these skulls, found in prehistoric Maltese temples at Taxien, Ggantja and Hal Saflieni, were displayed in the Archaeological Museum of the Valletta. A few years ago, though, they were removed and placed in storage. The public has not seen them since. Only the photographs taken by Maltese researcher Dr. Anton Mifsud and his colleague Dr. Charles Savona Ventura remain to testify of their existence and as proof of their extraordinary characteristics. Books by the two doctors, who helped us during our stay, provided the necessary documentation.

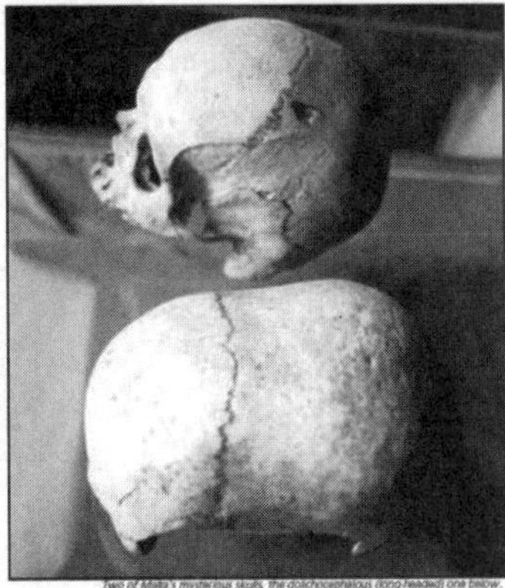

Two of Malta's mysterious skulls, the dolichocephalous (long-headed) one below.

Illustrations in the books show a collection of skulls with unmistakable abnormalities and/or pathologies. Some revealed nonexistent cranial knitting lines, abnormally developed temporal partitions, and drilled and swollen occiputs (the back, or posterior, part of the head or skull; the region of the occipital bone), as might appear after trauma recovery. Most significantly, however, they included an odd lengthened skull, bigger and stranger than the others, lacking the median knitting. Such a finding leads to several possible hypotheses. A similarity with skulls from Egypt and South America found with this particular deformity—ancient (from approximately 3000 B.C.), yet unique in medical pathology—suggests this could be an extraordinary discovery. Was this skull the result of an ancient genetic mutation between different races or something else?

The Long-Headed Skulls

Our request for help from the museum's management would certainly have failed if not for the intercession of the minister of tourism, Michael Refalo, who we met at the end of a press conference organized by Dr. Robert Zammit, of the Maltese Provincial Tourism Board. The minister, whose interest in the matter we stimulated, accompanied us to the museum, and personally obtained the director's permission to bring out the mysterious skulls, kept for about 15 years far from the sight of curious onlookers and researchers.

First page of an article on the elongated Maltese skulls in *Atlantis Rising* magazine.

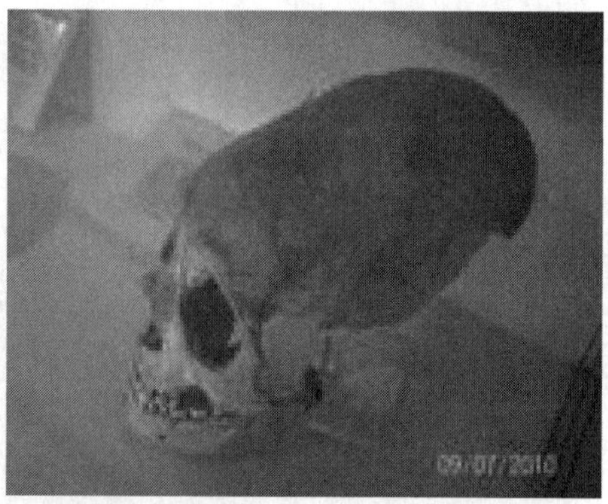

An elongated skull on display in the museum of Tuchersfeld, Germany.

It has been noted by researchers that the Hungarians and Huns also practiced cranial deformation as shown in these old paintings.

CHAPTER 6

CRANIAL DEFORMATION IN ASIA & THE PACIFIC

We elongate the heads of our children because it us our tradition and it originates with the basic spiritual beliefs of our people. We also see that those with elongated heads are more beautiful, and such long heads also indicate wisdom.

—South Malakulan saying translated by Kirk Huffman
for the Australian Museum

Taoist Coneheads of Ancient China

It is known that headbinding, as well as footbinding, was practiced in ancient China and Korea. Though records are sketchy, the ancient Chinese were known to have popularized a number of deformation techniques, including these and the creation of eunuchs. However, not many actual skulls have been found in China that are dolichocephalic, probably due to the moist acidic soil and crowded conditions prevalent in most of the country. Taoist art does show a number of Taoist "masters" and "immortals" with extended crania and even the curious "M" skulls that are also featured in some of the Olmec art. Some of the depictions of these ancient Taoist practitioners are indeed startling in their portrayal of cranial deformation. Most important of the Taoist immortals with an elongated head is Shou Xin Gong (or variably, Shou Xing) whose name translates as "Star of Longevity."

The Taoist immortals are considered "patron saints" of the Taoist belief system. Images of them can be found in porcelain, wood, ivory and metal reproductions as well as in paintings. Examples of writings about them include such Yuan and Ming Dynasty classics as *The Yueyang Mansion* by Ma Zhiyuan, *The*

Bamboo-leaved Boat by Fan Zi'an, *The Willow in the South of the City* by Gu Zijing, and *The Eight Immortals Depart and Travel to the East* by Wu Yuantai.

They were representative of typical individuals and represented wealth and poverty, old age and youth, male and female. The Chinese believed that average human beings could, through hard study, learn the secrets of nature and become immortal. Normally, only eight immortals are officially recognized, but legends describe several more. These immortals were idolized and respected for their wisdom, humor, and moral lessons and became legends that almost everyone common person was intimately aware of. When the ten immortals are listed, Shou Xin Gong is the tenth and last in the list.

The ten Taoist immortals are usually listed in this way: (1) Cao Guo Jiu; (2) He Xian Gu; (3) Lan Cai He; (4) Han Xiang; (5) Zhang Guo Lao; (6) Lu Dong Bing; (7) Han Zhong Li; (8) Tie Gwai Li; (9) Mang Mu Nian Niang; (10) Shou Xing (or Shou Xin Gong). Because of his elongated cranium, we will focus on the story of Shou Xin Gong.

Shou Xin Gong—Conehead God of Longevity

In popular Chinese lore, Shou Xin Gong is both an immortal and a powerful symbol of long life. People often display his portrait on their birthday for good luck. He is perhaps the most recognized of the immortals because his likeness is used for so many purposes. You can find statues of him and his wife in Chinese gift stores, and pictures of him appear on posters, gift wrap and gift cards, especially those meant for birthdays. The Chinese, who are very superstitious and have a real conviction in the concept of "luck," find Shou Xin Gong to be a powerful good luck icon.

As God of Longevity, Shou Xing is the third god of the Gods of Good Fortune. He is associated with the "Star of Longevity," or the Star of the South Pole, which occupied a very important position in star worship. The worship of stars retained its importance in China's royal ceremonies throughout the ages, even though dynasties rose and fell. People strongly believed that the stellar deity could control the fate of the country and, in addition, possessed the power to decide the life span of mortals. At some period in time, the stellar deity was endowed with human qualities and Shou Xing eventually become known as Nan-ji-Xian-weng - "The Old Man of the South Pole."

This is one of the most recognizable deities in Chinese literature. The God of Longevity is elderly, with a long wispy beard and an enormous, high bald head. In one hand, he always carries a long knotted staff and a pumpkin gourd, which holds the water of life. In the other hand, he holds the peach of immortality; sometimes there is a crane on top. Both items are symbols of immortality. Symbolically he is represented as a mushroom or a turtle.

A detailed folktale has grown up around Shou Xing. As it starts out, a young woman stood outside her house one night and looked up at the sky. She was surprised to see a very bright star and she pointed it out to her husband. Her husband told her that it was the Star of the South Pole and on that night the star was brighter than usual.

As a neighbor (who had overheard the conversation) came out to see, the Star of the South Pole suddenly disappeared. At that precise moment, the young woman began to feel unwell and so the couple retired for the evening. In the middle of the night, the young woman was awakened by the sound of a young boy. He cried hysterically and told her that she was his mother and he was in her womb, but she had to wait for ten years before he could be born. The young woman was terrified and woke her husband. She told him what she heard and then it struck her: could she be carrying the Star of the South Pole? They went outside to find the star but it was still gone.

Nine years later, the young woman was showing all the signs of pregnancy, and feeling that this was an unbearable wait. One day, she was being reflective and asked her son when he would come out. Suddenly, she felt something happening and heard her son telling her that it wasn't time yet. He said that in the tenth year, when the eyes of the stone dragon turned red, he would be born. So, it would be another year before their child would come! The neighbors heard this story and suggested that the husband paint the stone dragon's eyes red with pig's blood. The trick worked and their son was born after nine years of pregnancy. The parents named him Shou Xing, since the mother became pregnant when she looked at the Star of the South Pole.

However, to everyone's surprise, Shou Xing was totally bald. This was because his mother did not carry him to full term. As Shou Xing grew up, unfortunately, his hair did not grow. Because of his bald and elongated head, Shou Xing was embarrassed to go out in public and often went to the hills and practiced meditation and *The Way* (of the Tao). He sensed that something was calling him there every day. Eventually, Shou Xing decided to travel deep into the hills, where no one else had dared to travel. He had a calling. People asked him where he was going, and he replied he was going to the hills and would come back when it was time.

After 1000 years and nine generations later, Shou Xing finally returned to the village. He now had a long beard that was down to his waist. In his left hand he had a walking stick with a pumpkin gourd tied to it. In the right hand, he had the peach of immortality. By now it was the era of Jiayou during the Song Dynasty. Shou Xing went to Bianliang.

His strange, elongated bald head drew many people's attention. When the officials heard about him, they told the Emperor Renzong. The Emperor was curious to know more about this old man and thus, Shou Xing was brought before the Emperor. The Emperor asked him who he was and how old he was. Shou Xing did not answer and asked for some dates and wine to drink. After the old man drank nearly seven glasses of the Emperor's best wine, saying not a word, the Emperor was unimpressed and sent him on his way. Shou Xing left the palace and took a path out of town.

On the next day the official court astronomer came to see the Emperor. He told him that he saw the Star of Longevity descend down to the throne, and that the old man he saw yesterday could have been the stellar God of Longevity. The Emperor was furious that he did not get to ask Shou Xing the secret of long life, and ordered his men to bring him back to the court. However, Shou Xing had vanished and could not be found.

Then a local man reported that he saw Shou Xing walking on a country path in front of him, and as he turned the corner he was stunned to see Shou Xing ascend to the heavens in a cloud of smoke. The God of Longevity had appeared in mortal form, and from this time on, there were many Chinese stories about the Old Man of the South Pole bestowing more years on mortals.

This classic story is familiar to most Chinese, along with the many representations of Shou Xing as he is described in the story. Most commentators say his bald elongated head symbolizes his tremendous knowledge. Some pertinent questions would be: Was the "South Pole Star" which shone so brightly and then abruptly disappeared a UFO? Had Shou Xing's mother been impregnated by an extraterrestrial? Was this why he had an unusually elongated and bald head? Did he depart in a UFO at the end of his very long life? These are fascinating matters for speculation, but we will probably never know the answers.

In the alternative, was Shou Xing one of the ancient ones of the Shang Dynasty who practiced cranial deformation, used magic mushrooms and traveled across the Pacific to find the "Center of the World" and interact with the Olmecs? The Olmecs also had elongated heads, took hallucinogenic mushrooms, valued jade and even had the curious practice of creating eunuchs. Connections between ancient China and the Olmecs are very strong, as is shown in the David Childress book *The Mystery of the Olmecs*. Since cranial deformation was very common with the Olmecs, we would presume that it existed among the Shang Chinese, although specific evidence in archeology and art is not prolific as it is with the Olmecs. Both cultures thrived circa 1200 to 1500 BC. This is also the time period of the elongated Atonist skulls in Egypt.

Will it be shown that many of the elongated skulls from Paracas in Peru are also from this time period? Perhaps in the story of Shou Xing it possible that the "Star of the South" refers to a land in the southern hemisphere such as Peru. Indeed, Peru is a southern land of coneheads, gold and immortality-inducing (so the Chinese believed) psychedelic drugs such as the San Pedro cactus, which some researchers believe the Candlestick of the Andes at Paracas represents.

The Strange "M" Skulls of Ancient China

As noted above, cranial deformation originated by at least 1500 BC, with both the Olmecs and the Mitannis engaging in headbinding at that time. Was it being used in China at this time as well? This would appear to be the case.

Because the soil in China is generally wet and acidic, skeletal material does not survive long and we have few ancient skulls from China. When one searches books or the Internet for articles about strange crania being unearthed in China, there is virtually nothing written on the subject. Yet, it is known that art starting in 500 BC and continuing up until the 19th century AD depicted Taoist masters and immortals as often having elongated skulls and strange "M" shaped skulls. (See some of the illustrations in this book.) Though Taoism as we know it today began around 500 BC, Taoists texts themselves constantly refer to "ancient masters" who were wise and very knowledgeable about all things. Did these ancient ones have oddly-shaped heads?

The strange "M" shaped skulls are also seen in Olmec art and statuary, but few examples of this type of skull have been found. Very few Olmec skulls have been discovered, because wet acidic soil similar to China's also makes up both coasts of Mexico in the Isthmus of Tehuantepec where the Olmecs lived. One "M" shaped skull can be seen at the Archeological Museum at Merida in Mexico's Yucatan Peninsula. This skull is probably a Mayan skull from around 800 AD, but it is important to note that the Mayans inherited most of their customs from the Olmecs, including headbinding and other deformations such as body piercing and tattooing.

Body piercing and tattooing, having regained great popularity today, are historically related to cranial deformation, and one can generally assume that many individuals that had elongated or otherwise deformed crania also had various tattoos and body piercings, including ear, nose and lip rings. The subject of tattooing is beyond the scope of this book, but it was known to have been practiced by a wide array of peoples, from Egyptian and Arabian Bedouins to Pacific Islanders, Celtic tribes and the Olmecs, among others.

Cranial Deformation on the Korean Penninsula

With a little investigation on the Internet, it can be found that cranial deformation was practiced in ancient Korea. The Pusan

National University Museum on South Korea's east coast has a display and webpage on the deformed skulls from the Samguk Period in Korea's history (100 BC-668 AD).

Says the website:

> Owing to the soil characteristics of Korea, ancient human skeletons hardly remain and so physical anthropological studies about our ancestors have not been developed. However, many well-preserved ancient human skeletons were excavated from the Yean-ri site in Gimhae and the Neuk-do site in Samcheonpo by our museum, and these specimens furnished precious data for the relevant studies. Especially, through the study of the ancient skeletons from the Yean-ri site, the physical characteristics of the Gaya people came to be understood: according to this study, the Yean-ri men and women were 164.7 cm and 150.8 cm tall respectively on the average, and the dead under 12 years old accounted for one third of the total, implying the high mortality of infant and child at that time.
>
> In addition, it was elucidated that about 30% of the Yean-ri women buried in the early 4th century site had the indications of *pyeondu* ("frontal cranial deformation"). The *pyeondu* custom is a kind of cranial deformation custom to make a deformed head shape by pressing the forehead with a wooden board or stone since infancy. It is known that this custom was a unique custom of Byeonhan and Jinhan because it also corresponds to the article about Byeonhan and Jinhan in *Sanguozhi Weishu Dongyizhuan*: it says that "all the Jinhan people have a deformed skull as before because their heads were pressed with stones during their infancy."

So, we know that the Koreans were doing cranial deformation circa 100 BC, but could it have been practiced even earlier? We know that cranial deformation was being practiced in Egypt, Mitanni and Mexico circa 1200 BC. Lapita pottery discovered on Pacific Islands has been dated to 800 BC and earlier. Some Pacific islanders practiced cranial deformation to this day as we are about to see. The question is, was cranial deformation going on in Korea (and possibly Japan) as far back as 800 or 1200 BC?

Headbinding in Vanuatu

Far across the world, in what is now called Melanesia, we find examples of an elongated head presence and headbinding tradition. Melanesia is a sub region of so-called Oceania extending from the western end of the Pacific Ocean to the Arafura Sea, and eastward to Fiji. The region comprises most of the islands immediately north and northeast of Australia. The name *Melanesia* (from Greek: μέλας *black*; νῆσος, *islands*) was first used by Jules Dumont d'Urville in 1832 to denote an ethnic and geographical grouping of islands distinct from Polynesia ("many islands") and Micronesia ("small islands.")

The Republic of Vanuatu is an island nation, a remnant of French colonization. It is an archipelago that is of volcanic origin, some 1000 miles east of northern Australia, 500 kilometers northeast of New Caledonia, west of Fiji, and southeast of the Solomon Islands, near New Guinea. Where the Vanuatu people came from, and indeed the Melanesian people as a whole, is the subject of much debate.

Headbinding continued in Vanuatu up until modern times, and is apparently still occurring, according to the Australian Museum's website. This brief site includes several interesting photos, including some of living people with elongated heads. Says the site:

> The Culture Hero Ambat is associated with the origin of head binding in certain coastal areas of southern Malakula, Vanuatu. Ambat himself had an elongated head and a fine, long nose. Head elongation styles vary slightly among the many different language and cultural areas of southern Malakula. The area where people have the longest elongated heads is the Nahai-speaking area of Tomman Island and the southwestern Malakulan mainland opposite. A person with a finely elongated head is thought to be more intelligent, of higher status and close to the world of the spirits. Even today, throughout Vanuatu, the

> Bislama/Pidgin English term, Longfala Hed (Long Head) is synonymous with intelligence.
>
> On Tomman Island and the facing south-southwestern Malakula mainland, headbinding began approximately a month after birth. Each day the child's head was smeared with burnt paste made from the Navanai-Molo nut (from the candle nut tree). This process softens the skin and prevents 'binding rash'. The child's head was then bound with Ne'Enbobosit, a soft bandage made from the inner bark of a type of banana tree. Over this was placed a No'onbat'ar (specially woven basket) made from Nibirip (pandanus) and this was bound around with the Ne'euwver (fiber rope).

This process continued every day for approximately six months to produce the required shape. The reason that the tribesmen practiced cranial deformation was that they believed that long-headed people were more intelligent, apparently because of some early heroes with coneheads that they greatly admired. An interesting note about the cultural hero Ambat is that he and his children were white, with long straight hair. In a story paralleling that of the Biblical Adam and Eve, the children one day ate a rose apple, which their father had forbidden them to do. As punishment, he turned them black and banished them to the south of the island (Malakula) where they were to wear *nambas*, or penis sheaths. When the explorer Captain Cook arrived at the island in 1774, he was given a hero's welcome in honor of the white god Ambat. When the pale-faced slave traders started coming around, however, relations quickly deteriorated, and ships were often attacked and the crews eaten.

The Australia Museum quotes one Vanuatu native as saying about headbinding:

> "We elongate the heads of our children because it is our tradition and it originates with the basic spiritual beliefs of our people. We also see that those with elongated heads are more handsome or beautiful, and such long heads also

> indicate wisdom." *(General South Malakulan Quotation as translated into Bislama by Kirk Huffman for the Australian Museum)*

In fact, headbinding is still occurring on Vanuatu and a brief video can be seen on YouTube entitled, "Malekula String Band and the Longheads" which was uploaded by "selfshootingPD" on February 12, 2007. It is a video of a string band playing to a small crowd on a beach on the island of Malekula and in the audience are several "longheads" from the Nambas tribe. The camera closes in on several persons who clearly have elongated skulls and are very much alive. One is holding a baby who has had its cranium elongated.

The video footage was apparently shot in 2004 and the people who posted it included this brief text to go along with their video:

> They are the real life coneheads. In 2004, on a remote island in the South Pacific, exclusive footage was shot of 'headbinding' or skull elongation which the *Lonely Planet* said had "...long since died out". On what became an impromptu, largely unplanned adventure-filled 'expedition' the filmmakers found that not only were these 'longfala hed' Small Nambas tribe members alive and well, but contrary to *Lonely Planet*'s assertion they were binding the heads of new-born babies—in time-honored fashion. Led by their 'progressive' chief, they had moved out of the dense jungle to the shoreline just weeks earlier. In this clip "longfala heds" listen to an island string band...

So we can see that headbinding has not died out entirely on Vanuatu (despite what the *Lonely Planet* guidebook has to say) and living coneheads in action can be viewed in the video. Malekula would seem to be a good place to go for any researcher wanting to do in-depth research into living people with elongated heads. In fact, the material posted with the video includes an article from the *Sydney Morning Herald* (27 Oct 2004) that says an anthropologist named Kirk Huffman has studied the islanders for 20 years.

According to him, "There's also a belief that head beautification increases your memory. ...If it's done properly, there are no side effects."

The Australian Museum also says that among the Bintulu Malanus Dayak people of Borneo it was considered a sign of beauty to have a flat forehead. The process of flattening the head was started during the first month of a baby's life with a tool called a *tadal*. A cushion was placed on the child's forehead with bands placed over the top of the head and around the back of the head. The strings holding the bands in position were adjusted without disturbing the baby. In the early stages, only very slight pressure was applied, but gradually the pressure was increased.

It has been pointed out by a number of archeologists, including myself, that the famous statues of Easter Island essentially depict men who have elongated heads. Yes, the people depicted in these statues are dolichocephalus. They have high, flat foreheads as well as narrow, flat backs of the head. They are "flatheads" in conehead vernacular.

Was cranial deformation popular throughout the Pacific at some point? Were transoceanic voyagers starting in southern Asia and voyaging across the many islands of the Pacific to the Americas bringing cranial deformation with them? Perhaps Vanuatu and Easter Island are the last bits of evidence we have for a Pan-Pacific distribution of cranial deformation to Mexico and Peru.

Cranial Deformation in the Pacific Northwest

We have already discussed how the Olmecs and the Mayas of Central America had dolichocephalous crania in many cases. Though skeletal material is often difficult to find in tropical areas because of the fast rate of decomposition, many jade figurines attributed to the Olmecs have been found which show persons with elongated heads, so we know that this curious "fashion" was spread from the Andes to the jungles of Panama, Costa Rica and Nicaragua to the coasts and mountains of Mexico. It even reached the Pacific Northwest of the United States.

The largest freshwater lake west of the Mississippi is Flathead Lake in western Montana. The lake was given this name because the local Native Americans in the area were called the Flathead

Indians. They were called the Flathead Indians because they had the curious elongated and flattened heads that we now know as dolichocephalous crania.

Similarly, at the time the first European explorers got into the Seattle area of the Pacific Northwest, they found that the local Chinook tribes had elongated and flattened skulls. The early Russian, Spanish and English explorers who came into the area documented this strange custom, and early drawings exist of a woman with a dolichocephalous cranium who is holding a child in a deformation cradle on her lap. The child will have an elongated head much like his mother through this headbinding technique. Curiously, she appears in the drawing to have unusually small feet, possibly bound when she was an infant, as well as writing-like tattoos on her skin.

How widespread cranial deformation was in the Pacific Northwest is not known. This rainy area of North America does not preserve skeletal material very well and few tombs or other burials have been discovered. A very startling discovery was allegedly made in the Aleutian Islands of Alaska during WWII, and we will discuss this odd incident in the next chapter.

One of the Taoist immortals, Shou Xin Gong, usually depicted with an extremely long cranium.

Left: One of the Taoist immortals, Shou Xin Gong, usually depicted with an extremely long cranium. *Below*: Drawings of the Eight Taoist Immortals.

A depiction of three of the Taoist immortals with the two on the right having the curious "M" cranium.

A depiction of one of the Taoist immortals with the "M" cranium.

Taoist immortals, depicted as winged birdmen, walking on clouds and holding zhi ("numinous mushrooms") in their hands. (Han Dynasty stone reliefs from Nanyang, Hunan).

A depiction of one of the Taoist immortals with a pronounced "M" cranium.

Above: A painting of the Taoist immortal Quan Yin. She rides a dragon and has a long, narrow headress. *Opposite page*: One of the Taoist immortals, Zongli Quan, depicted with a curious bump on his forehead.

A statue of Taoism founder Lao Tzu located in Quanzhou, China.

One of the Taoist immortals, Lu Dong Bin, depicted with a high forehead and his hair in a top-knot.

Russian skull with elongated crania.

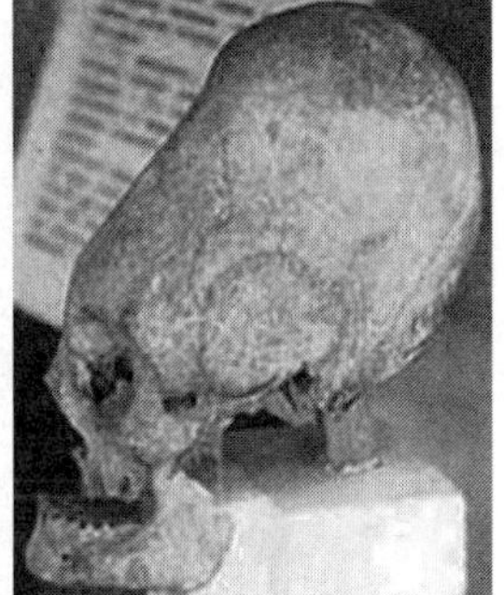

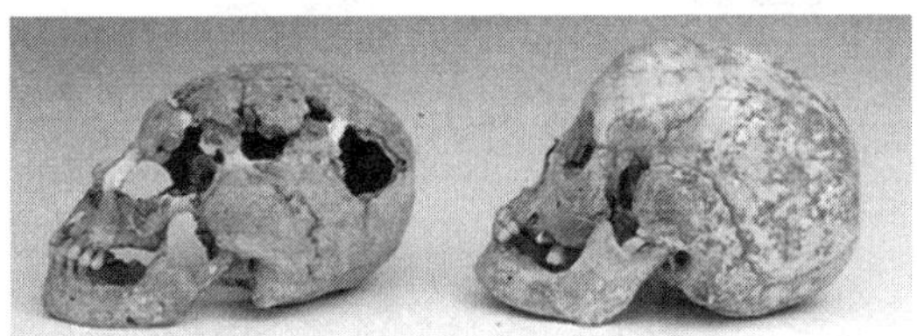

Korean Yean-ri skulls showing their elongated crania.

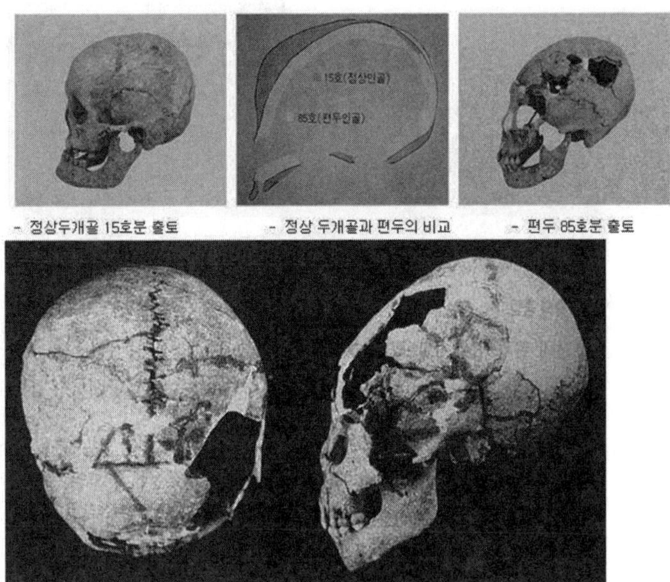

- 정수리~후두부가 터진 5~6세 소아 두개골

Korean Jinhan skulls showing their elongated crania.

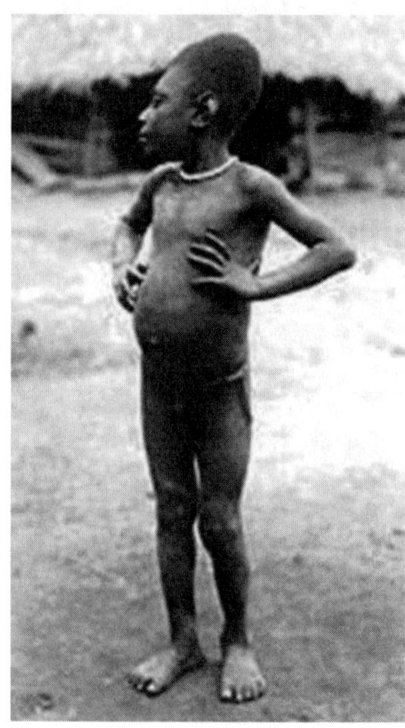

Above: Two photos from the Mangbetu tribe, circa 1940. Compare to the boy from Vanuatu, below. *Left*: A photo, circa 1940, of a young boy from Vanuatu with an extended cranium.

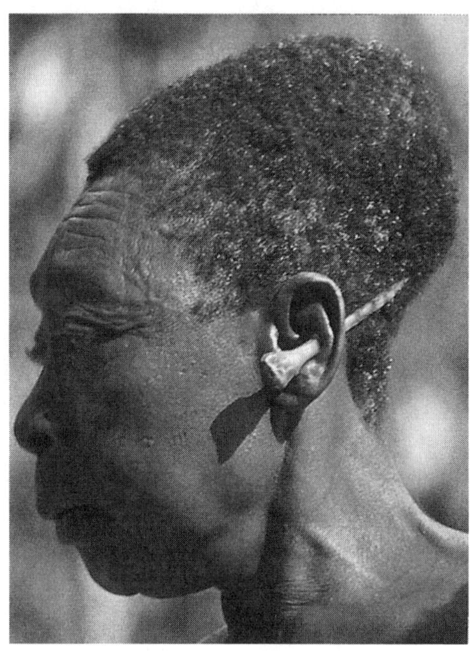

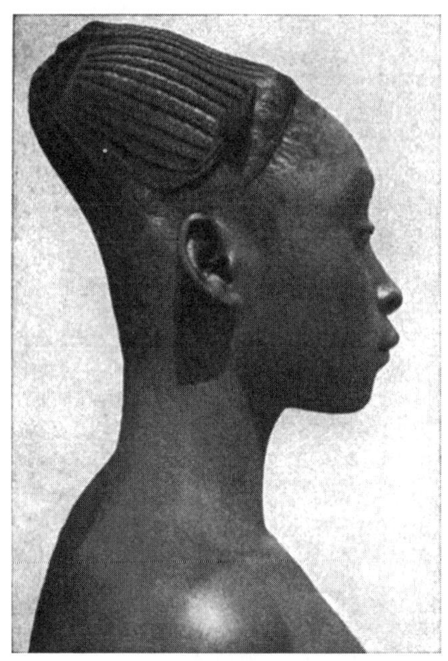

Above: A photo, circa 1970, of a woman from Vanuatu with an extended cranium, known to be the result of headbinding. *Left*: Compare her head with the head of a Mangbetu woman, circa 1940.

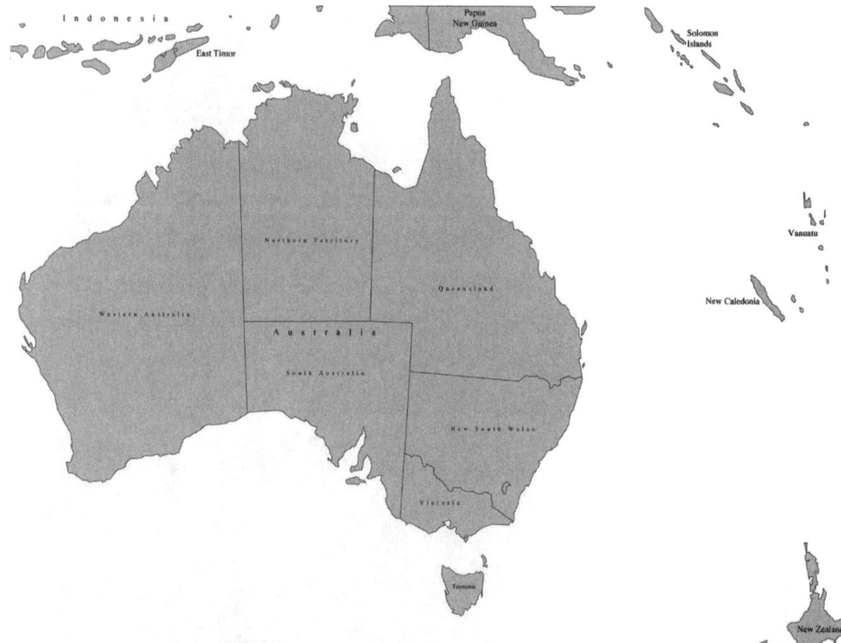

A map showing the position of Vanuatu in Melanesia.

A picture of a Flathead Indian of Montana painted in 1851.

A drawing of a Chinook woman with an elongated head and a cradle for forming the head of her infant.

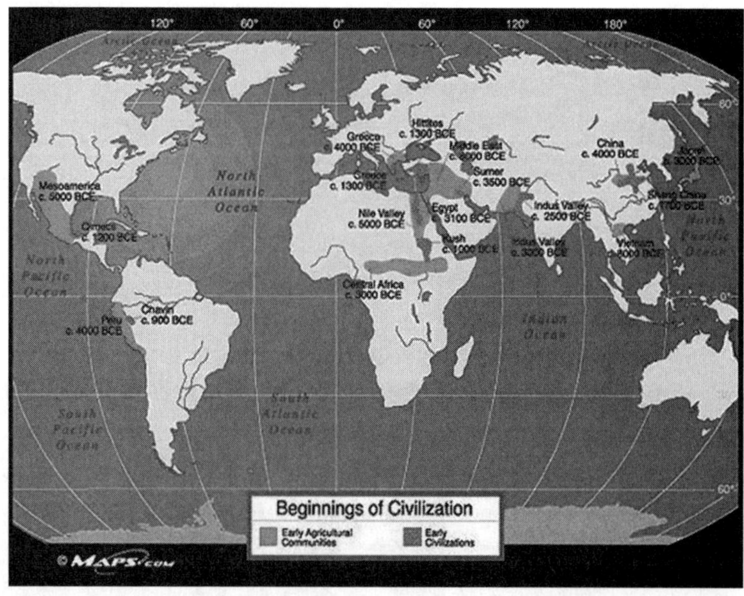
Beginnings of Civilization
Early Agricultural Communities
Early Civilizations
North Atlantic Ocean
South Atlantic Ocean
North Pacific Ocean
South Pacific Ocean
Indian Ocean
Mesoamerica c. 5000 BCE
Olmecs c. 1200 BCE
Chavin c. 900 BCE
Peru c. 4000 BCE
Greece c. 4000 BCE
Greece c. 1300 BCE
Hittites c. 1300 BCE
Middle East c. 8000 BCE
Sumer c. 3500 BCE
Egypt c. 3100 BCE
Nile Valley c. 5000 BCE
Kush c. 1000 BCE
Central Africa c. 3000 BCE
Indus Valley c. 2500 BCE
Indus Valley c. 3200 BCE
China c. 4000 BCE
Vietnam c. 2000 BCE
© MAPS.com

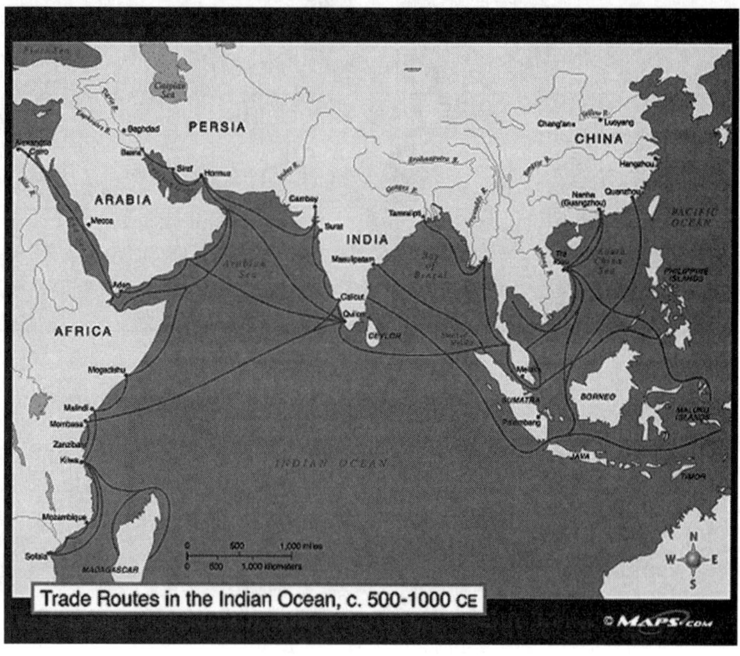
PERSIA
ARABIA
INDIA
CHINA
AFRICA
Baghdad
Mecca
Aden
Hormuz
Siraf
Cambay
Surat
Tamralipti
Masulipatam
Calicut
Quilon
CEYLON
Arabian Sea
Bay of Bengal
INDIAN OCEAN
PACIFIC OCEAN
SUMATRA
BORNEO
JAVA
TIMOR
Palembang
Melaka
Chang'an
Luoyang
Hangzhou
Quanzhou
Nanhai (Guangzhou)
PHILIPPINE ISLANDS
MOLUCCA ISLANDS
Mogadishu
Malindi
Mombasa
Zanzibar
Kilwa
Mozambique
Sofala
MADAGASCAR
0 500 1,000 miles
0 500 1,000 kilometers
N W E S
Trade Routes in the Indian Ocean, c. 500-1000 CE
© MAPS.com

DESERT
Hwang R.
present bed
Great Wall
Chang Yeh
Tai Yuan
An-ting
THE LOESS REGION
Lung Hsi
Wei R.
Chang-an
Honan
Loyang
Hanchung
CHINESE PLAINS
Present bed of Hwang R.
Ching-chow
SHANTUNG
Tung-hai
Huai (Hwang) R.
Kuang-ling
Yangtze R.
Wuchun
Hui-hi
THE RED BASIN
Kwang Han
Yangtze R.
Chiang Hsia
Lu-chiang
CHINESE PLATEAU
Ching-chow
Yu-chang
THE HILL COUNTRY
Min Yueh
Si R.
Tsang Wu
Hanoi
TAIWAN
ANCIENT CHINA
Scale 0 300 Miles
Raisz

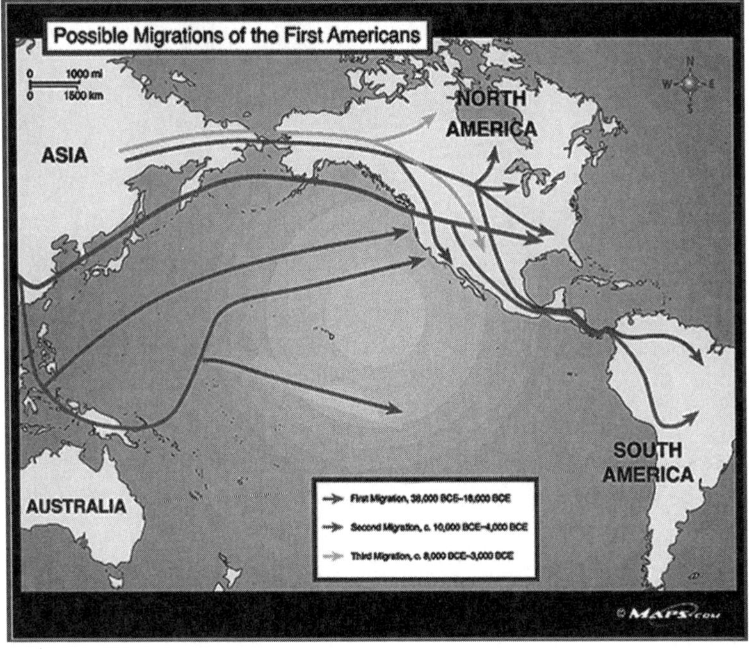
Possible Migrations of the First Americans
0 1000 mi
0 1500 km
N
W E
S
ASIA
NORTH AMERICA
SOUTH AMERICA
AUSTRALIA
First Migration, 38,000 BCE–18,000 BCE
Second Migration, c. 10,000 BCE–4,000 BCE
Third Migration, c. 8,000 BCE–3,000 BCE
© MAPS.com

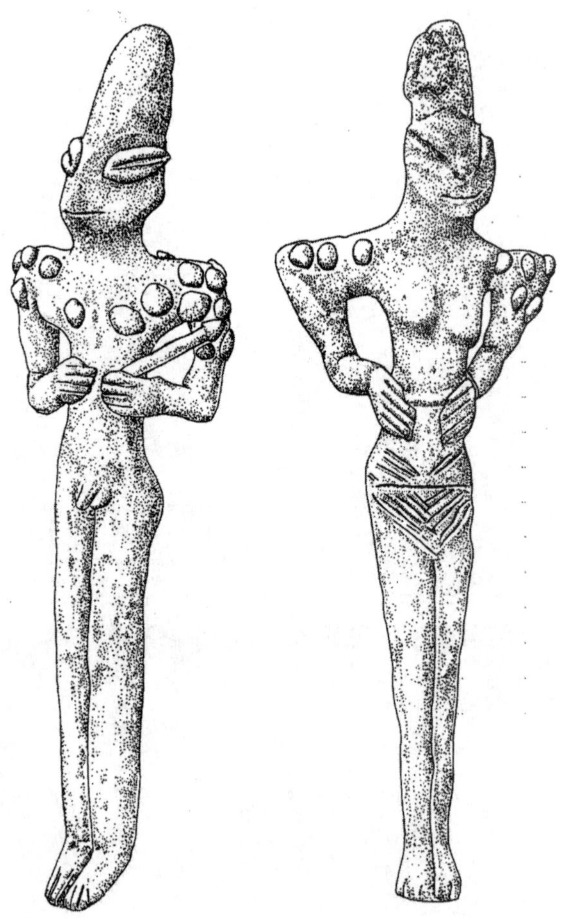

Two Sumerian clay figurines from the Ubaid Period (5000 BC) showing elongated heads. The male on the left is from Eridu and the female on the right is from the ancient city of Ur.

CHAPTER 7

THE NEFILIM, THE WATCHERS AND ELONGATED HEADS

This isn't the Mitchell-Hedges skull,
note the elongated cranium.
—Indiana Jones and the Crystal Skulls

The strange world of cranial deformation gets more bizarre the more one researches the subject. With people all over the world making use of cranial deformation–even the famous *moai* statures of Easter Island are coneheads—the big question is why. Where did the ancient civilizations that practiced this artificial skull deformation get the idea? No one seems to have a good answer.

The standard explanation is that it was a style of the day and the "technology" of deforming crania was relatively simple; it could be done by any family that had a newborn infant, a couple of pieces of wood and some rope. All it really took was some patience and time. But once again we have to ask ourselves—why would this be the style? It seems pretty odd. Also, given the currently reigning isolationist doctrine of mainstream archeology, this strange style adopted by widely separated cultures would have to have developed independently in each place. This seems almost impossible. It would seem that the observed widespread participation in such a curious fad would require a certain diffusion of ideas.

The idea that there was some other race of people, possibly extraterrestrials, who naturally looked this way without any artificial help is a shocking one. This idea is never discussed in academic journals, and the literature on the subject is sketchy. Perhaps the most prominent promoter of such a theory was the late Zecharia Sitchin in his books on the so-called 12th Planet and the ancient alien "Anunnaki" race who interacted with the Sumerian and other cultures (according to him).

Were practitioners of cranial deformation desirous of looking like a race of beings that actually had elongated skulls? The best hypothesis in our opinion is that the headbinding seen worldwide was done in emulation of an ancient long-headed race (either natural or forced) that brought wisdom and knowledge to our ancestors. Some fairly direct evidence in favor of this hypothesis is that the Nambas tribe of Vanuatu, who still practice headbinding today, think it makes people more intelligent and do it in honor of Ambat, an ancient long-headed cultural hero who has come to be worshipped as a god.

An Island Graveyard of Coneheads

In his 1969 book *More "Things"* zoologist and author Ivan T. Sanderson discusses a letter he received from an engineer who was stationed on the Aleutian island of Shemya during World War Two. While building an airstrip, his crew bulldozed a group of hills, discovering under several sedimentary layers what appeared to be a graveyard of seemingly human remains, consisting of crania and long leg bones. The crania measured from 22 to 24 inches from base to crown. Such a large cranium would imply an immense size for a normally proportioned human, about twice the normal size. Furthermore, every skull was said to have been neatly trepanned!

Sanderson tried to gather further proof, eventually receiving a letter from another member of the unit who confirmed the report. The letters both indicated that the Smithsonian Institution had collected the remains, yet nothing further was heard. Sanderson seems to be convinced that it was not a hoax, but wondered why the Smithsonian would not release the data. To quote him, "...is it that these people cannot face rewriting all the text books?"

The details are sketchy, but one wonders if the skulls have been elongated, meaning they would be out of proportion to the rest of the remains. This island of gigantic (elongated?) skulls may have been a stopover point for a mysterious group of coneheads who were coming from Asia in boats and going along the Pacific coasts of North and South America to spread their culture, technology and trade routes. Was Shemya Island a giant graveyard for the Anunnaki or Taoist immortals?

In the 1880s, a number of strange discoveries in Minnesota mounds were published in the *Minnesota Geological Survey,* the St. Paul *Pioneer* and the St. Paul *Globe*. At Chatfield, Minnesota, six skeletons of enormous size were discovered. Similarly, at Clearwater, Minnesota, seven gigantic skeletons were uncovered in mounds. They were reported to be buried head down, and their skulls had receding foreheads and complete double dentition (two rows of teeth). The receding foreheads are characteristic of the "flatheads" and "coneheads."

The Coming of the Coneheads to Malta

Karen Mutton, in her book *Scattered Skeletons in Our Closet*, discusses Italian writer Adriano Forgione's research on the skulls discovered in the Maltese temple tombs at Tarxien, Ggantja and Hal Saflieni. He was only able to study the skulls at the Maltese Archaeological Museum of Valletta when Michael Refalo, the Minister for Tourism, accompanied him to the museum and obtained the director's cooperation. The skulls had been removed from display in 1985.

Forgione found one skull in particular to be highly dolichocephalic, "bigger and more peculiar than the others, lacking of the median knitting or suture." Other skulls were more natural in appearance yet "still presented pronounced, natural dolichocephalous" shapes "distinctive of an actual race." Many of the 7,000 skeletons dug out of the Hypogeum exhibited "artificially performed deformities."

So Forgione thinks that a number of the remains found are those of a separate race with naturally elongated heads. You may recall that Maltese archaeologists Dr. Mark Mifsud and Anthony Buonanno concurred with this idea, telling Forgione, "They are another race although C14 or DNA exams haven't yet been performed."

Forgione formulated a theory that the skulls originated with a race that settled in Mesopotamia about 7,000 years ago. In 1933 Max Mallowan excavated Neolithic graves at Tell Arpachiyah in Iraq dated from 4600 BC to 4300 BC (from the Halaf and Ubaid periods). He reported the discovery of skulls having a "marked degree of deliberate, artificial deformation," leading to an "elongated skull." A 1996 monograph on Mallowan's discoveries by Stuart Campbell remarked, "Skull deformation at Arpachiyah appears, on current (1995) knowledge, striking... Skull deformation seems to occur with regularity at other sites of this general period over a very wide area... Jericho, Chalcolithic Byblos, Ganj Dareh, and Ali Kesh." Forgione, who (as we have seen) was very taken with the serpentine look that would result from cranial elongation, noted that in urbanized centers such as Jarmo, Iraq, the mother goddesses were represented as divinities with lengthened heads and the faces of vipers.

According to Forgione, this ancient race is also associated with the 'Nephilim' of Genesis; they eventually settled Egypt in pre-dynastic times. They dispersed to Malta in about 2500 BC, but still survived in Egypt a millennium later as evidenced by the mysterious pharaoh Akhenaton who was always depicted with an extremely elongated skull. This is largely the conclusion that British author Andrew Collins came to in his books, to be discussed shortly. The Egyptologist Walter Emery came to a similar conclusion in the 1930s when he discovered a large elongated skull at Saqqara.

Says Mutton:

> Dolichocephaloids also appeared in predynastic Egypt and in the art of the New Kingdom Armana period that covered the era of King Akhenaton. Professor Walter Emery excavated Saqquara in the 1930s and discovered amongst other skeletons a dolichocephalic skull larger than the others. He postulated it was from a different ethnic group that wasn't indigenous to Egypt but had performed priestly and government roles. He associated them with Shemsu Hor, 'the disciples of Horus.'

Nephilim, Extraterrestrials and Atlantis

Mainstream archeologists have acknowledged in the last few years that the Olmecs and other cultures with elongated skulls are key elements in the "ancient astronaut" field of speculation. When Ignacio Bernal was publishing the first overviews of the Olmecs in the late 1960s, the ancient astronaut hypothesis had not really entered pop culture. But soon, the gigantic African-looking heads were part of the hypothesis that some of the ancient gods were actually astronauts. Authors such as George Hunt Williamson, Robert Charroux, Erich von Daniken, Zecharia Sitchin and others have connected the ancient Sumerians (and others) with the Olmecs and extraterrestrials.

The ancient astronaut theory largely maintains that "space aliens" came to our planet and helped shape ancient civilizations. These extraterrestrials had naturally elongated heads and oriental features (such as narrow, slanting eyes, often called by anthropologists "coffeebean eyes").

Statues and other figurines found at ancient Sumerian sites in Iraq indicate that some of the inhabitants of the area had elongated heads and the puffy, slit-like "coffeebean" eyes. Such figurines give the impression of a reptilian countenance, and are today called by archeologists "serpent priest" or "lizard" figurines. Were they humans or extraterrestrials?

In the extraterrestrial theory, the ETs were the overlords of the primitive Earth populations (which were possibly genetically engineered by them) and were highly admired by the humans who were their subjects. As these extraterrestrials returned to their own planet, the humans left in charge decided to emulate the aliens, who had elongated heads and looked distinctly different from normal humans. The ruling elite of the humans therefore began the practice of skull deformation in order to try to look like the extraterrestrial masters who had once ruled them.

A similar theory on the Olmecs and their practice of cranial deformation is that they were remnants of one of the outposts of the lost continent of Atlantis. In this theory, Atlanteans, for reasons unknown, liked to have long, conical shaped skulls, perhaps to enhance psychic power. Atlantis was composed of many different races, much like the United States, which would explain the variety of people depicted in the art of Olman.

It is thought that Atlanteans, in their worldwide journeys, impressed other cultures with their high level of civilization and knowledge of all things psychic and scientific. The people visited or colonized by the Atlanteans, such as those in Mexico, Peru and

Egypt, began to imitate the Atlanteans and their unusual customs such as headbinding.

This would help to explain why cranial deformation is so widespread. As we have seen, it can be found in cultures worldwide. This theory extends to the Tiahuanaco culture of Lake Titicaca on the Altiplano of Bolivia (who also practiced headbinding). It states that there were many remnants of the Atlantean civilization, all aligned with one another, sometimes called the Atlantean League. One of the hallmarks of the ancient Atlantean sea-kings is the wearing of the turban, which in itself is associated with headbinding. Are turbans the final remaining manifestation of the great Atlantean League?

It seems that one of the goals in headbinding was to make the cranium as long as possible. It made the person taller, gave him more brain capacity, and, needless to say, a very special look that would distinguish the conehead from other folks with normal crania. One has to wonder if the person with longest cranium was deemed the more important "priest," "priest-king," or whatnot. The more we look into cranial deformation, the more puzzling it becomes, and no simple answers for the phenomenon are forthcoming. This may be the reason most archeologists steer away from the subject as much as possible; it quickly leads into a quagmire of questions that are best left unanswered if the theory of isolationism is to be continued unchecked. How would such a practice have originated in the first place? Why would independent cultures—allegedly without contact with each other—have created such a practice? Were persons with elongated crania of a special royal class in each of these societies? Did ancient brain surgery, trepanning and modern medical practice originate as a result of cranial deformation?

Zecharia Sitchin and the Extraterrestrial Anunnaki

The strange looking goddess figures from Sumer seem to show cranial elongation, and were abundantly produced as works of art and probable worship. Zecharia Sitchin (July 11, 1920–October 9, 2010) devoted his life's work to the mysteries of Sumer, interpreting Sumerian mythology and symbolism in a way that is unique. A full analysis of his theories of an alien race, called the Anunnaki, who supposedly came to earth and genetically manipulated humankind, is beyond the scope of this book, but a short synopsis may be useful.

Sitchin was an Azeri-born American author of books promoting an explanation for human origins involving ancient astronauts. Sitchin attributes the creation of the ancient Sumerian culture to the Anunnaki, which he asserts was a race of extraterrestrials from a planet beyond Neptune called Nibiru. According to Sitchin's interpretation of Mesopotamian iconography and symbolism, outlined in his 1976 book *The 12th Planet* and its sequels, this undiscovered planet beyond Neptune follows a long, elliptical orbit, and reaches the inner solar system roughly every 3,600 years. According to Sitchin, Nibiru was called "the twelfth planet" because the Sumerians' conception of the solar system counted all eight planets, plus Pluto, the Sun and the Moon, so Nibiru would be the twelfth heavenly body in our solar system.

According to Sitchin, Nibiru collided catastrophically with another planet, once located between Mars and Jupiter, named Tiamat after a goddess in the Babylonian creation myth the *Enûma Eliš*. This and other collisions supposedly formed the planet Earth as we know it today, plus the asteroid belt and a number of comets. According to Sitchin, Tiamat broke in half when it was struck by one of Nibiru's moons, and then on a second pass Nibiru itself struck the two broken fragments. Half of Tiamat became the asteroid belt between Mars and Jupiter. The other half, struck again by one of Nibiru's moons, was pushed into a new orbit and became the planet Earth. Somehow, life persevered on the sphere throughout these great catastrophes, and the orbital year of the planet became the familiar 365 days that we know today.

Nibiru was the home of a technologically advanced human-like race called the Anunnaki. Sitchin wrote that they evolved after Nibiru entered the solar system, and they first arrived on Earth probably 450,000 years ago. They were looking for minerals, especially gold, which they found and mined in Africa. Sitchin maintains that these beings, that earthlings came to think of as "gods," were the rank-and-file workers of the colonial expedition to Earth from planet Nibiru.

The Anunnaki looked much like humans, he claims, except they had elongated skulls and were taller than normal humans.

Sitchin gives several illustrations of them, taken from Sumerian clay figurines, showing the Anunnaki as having coneheads and "coffeebean" eyes.

Sitchin writes in his books that the Sumerian "god" Enki suggested that primitive workers (*Homo sapiens*) be created through genetic engineering. By crossing extraterrestrial genes with those of *Homo erectus*, the Anunnaki could create slaves to replace their own workers in the mines. The Anunnaki gold miners had mutinied due to dissatisfaction with their working conditions.

So, Sitchin's extraterrestrial hypothesis has the human race being created as genetically manipulated apemen—a missing link similar to bigfoot or yetis. Sitchin maintains that ancient inscriptions report that the human civilization in Sumer was developed under the guidance of these visiting "gods," and human kingship was inaugurated to provide intermediaries between mankind and the Anunnaki (resulting in the concept of the "divine right of kings").

While there is widespread agreement among scholars that the ancient Sumerian texts refer to the Anunnaki as deities, Sitchin's interpretation that the Anunnaki are extraterrestrials is routinely scoffed at by most historians. Most doubt that there is an undiscovered planet beyond Neptune that follows a long, elliptical orbit, saying that such a planet would have been discovered already if it existed. Indeed, that human-type beings could live on a planet so far away from the Sun seems pretty fantastic. Perhaps they came from a closer and more hospitable planet—like Mars!

The mystery of the Anunnaki—who or what they were—is an enduring one, and Zecharia Sitchin has added quite a bit to the discussion. It would seem that the Anunnaki were indeed coneheads, but were they extraterrestrials? Could they alternatively have been an elite race of humans who lived in the remote mountains of northern Iraq, the land of the Kurds today? Possessing high technology such as the use of metals and large ships, could they also have used cranial deformation as a way to make themselves look different from other human beings? We will explore this possibility now.

The Watchers and the Anunnaki

With the Anunnaki-as-extraterrestrials theory we have the

concept that some human coneheads had their elongated skulls genetically encoded into their DNA—extraterrestrial DNA at that. If this theory were correct, we would surmise that there would be three different types of coneheads: (1) genuine from-another-planet extraterrestrials, (2) half human-half extraterrestrial beings, and (3) *Homo sapiens* who were artificially deforming their skulls to get the elongated effect.

An alternative and intriguing theory has been put forward by British writer Andrew Collins in his two books, *From the Ashes of Angels* and *Gods of Eden*. In these books, Collins proposes that the Anunnaki and the "Watchers" from *The Book of Enoch* (more about this later) are a special group of humans called "the Sons of God" who just happened to have elongated heads. They are descended from the mysterious Shemsu Hor ("Disciples of Horus") written about in early Egyptian texts.[94, 95]

You may remember that Karen Mutton mentioned the Shemsu Hor when discussing the famous Egyptologist Professor Walter B. Emery's find of a fair-haired dolichocephalic skull at Saqqara, Egypt. This ancient group is supposed to have been a sort of secret society dedicated to keeping the wisdom of the old solar religion. In his book *Archaic Egypt*, Emery writes:

> Towards the end of the IV millennium B.C. the people known as the Disciples of Horus appear as a highly dominant aristocracy that governed entire Egypt. The theory of the existence of this race is supported by the discovery in the pre-dynastic tombs, in the northern part of Higher Egypt, the anatomical remains of individuals with bigger skulls and builds than the native population, with so much difference to exclude any hypothetical common racial strain.

Collins says that ancient texts speak of the Watchers (who he believes to be descendants of the Shemsu Hor) as "angels" who lusted after mortal women who, subsequent to having relations with the Watchers, gave birth to giant offspring referred to as Nephilim. The texts also record how these apparently physical angels revealed to mankind the forbidden arts and sciences of heaven.

In his books, Collins shows that these human angels—described as tall, with white hair, burning eyes and viper-like faces—may well have originated in Egypt and been responsible for the construction of the Great Sphinx and the other megalithic monuments along the Nile shortly before the cataclysms that accompanied the end of the last Ice Age. This would be consistent with them being associated with the mysterious Shemsu Hor.

Assembling clues from archaeology, mythology and religion, Collins puts forward the hypothesis that Eden, 'heaven' and 'paradise' were all a single geographical area in Turkish Kurdistan. This same region was featured in Near Eastern mythologies as the abode of the gods, the realm of the immortals and the domain of angels. Here the Watchers would appear to have remained in virtual isolation before integrating with developing human societies on the plains below. The civilizations in Sumer, Akkad and Iran may all have developed as a result of the Watchers' influence on the people there.

Is the memory of interaction with the Watchers behind the stories of open contact between the gods and goddesses of Sumer and Akkad and mortal-kind? And what of the accounts of wars with bird-like demons, devils and vampires named Edimmu who lived in underworld domains beneath the sands of ancient Iraq?

Essentially, Collins maintains that civilization is the legacy of a race of human angels known as the Watchers, and these are the same as the Anunnaki. In support of his theory, Collins points out that many people see the Pentateuch (the first five books of the Old Testament) as littered with accounts of ethereal angels appearing to righteous patriarchs and visionary prophets. Yet this is simply not so, he says. For instance, the three angels who approach Abraham to announce the birth of a son named Izaac, and the two angels who visit Lot and his wife at Sodom prior to its destruction, are described simply as 'men'; they sit down with their hosts to take food like ordinary mortal people.

The Book of Enoch and the Watchers

The Book of Enoch was written in stages sometime before the time of Christ. It is called a pseudepigraphal work, meaning that it is attributed to, but not written by, Enoch. Although it was never included in the Bible, much of what is written in the New Testament, especially in the *Book of the Revelation*, draws from ideas and events first presented in its pages. The most interesting part of the story deals with the fall of a group of the angels known in the book as "the Watchers."

The Watchers were tasked with keeping an eye on the world, lending a helping hand from time to time, but they were never to interfere with the course of human events. At one point, a rogue group of 200 angels decide to interrelate with the human race in a very big way. They descended to the land of men and took "wives" with whom, as is mentioned above, they created a race of giants called the Nephilim, which in Hebrew means "those who have fallen." They also taught humankind the secret knowledge of the heavens. They taught the men the art of metalworking, and the making of arms and armor. They taught the women about beautification and jewelry, along with secrets of a sexual nature. They imparted wisdom pertaining to astronomy, geography, geology and meteorology. In short, they meddled in human affairs to a huge degree. Collins says they knew the possible repercussions they might face, as they swore an oath to the effect that their leader Shemyaza would "take the fall" if things went bad.

Collins comments on the situation:

In my opinion, this revelation of previously unknown knowledge and wisdom seems like the actions of a highly advanced race passing on some of its closely-guarded secrets to a less evolved culture still striving to understand the basic principles of life.

More disconcerting were the apparent actions of the now fully grown Nephilim, for it says:

> And when men could no longer sustain them, the giants turned against them and devoured mankind. And they began to sin against birds, and beasts, and reptiles, and fish, and to devour one another's flesh, and drink the blood. Then the earth laid accusation against the lawless ones.

By now the cries of desperation from mankind were being heard loud and clear by the angels, or Watchers, who had remained loyal to heaven. One by one they are appointed by God to proceed against the rebel Watchers and their offspring the Nephilim, who are described as 'the bastards and the reprobates, and the children of fornication'. The first leader, Shemyaza, is hung and bound upside down and his soul banished to become the stars of the constellation of Orion. The second leader, Azazel, is bound hand and foot, and cast for eternity into the darkness of a desert referred to as D–d*f*^l [not a recognizable word in English]. Upon him are placed 'rough and jagged rocks' and here he shall forever remain until the Day of Judgment when he will be 'cast into the fire' for his sins. For their part in the corruption of mankind, the rebel Watchers are forced to witness the slaughter of their own children before being cast into some kind of heavenly prison, seen as an 'abyss of fire.'[94]

Visage Like a Viper

Collins says that other Enochian texts have surfaced among the Dead Sea Scrolls which he thinks are very important. One of these texts is called the Testament of Amram. Says Collins:

> Amram was the father of the lawgiver Moses, although any biblical time-frame to this story is irrelevant. What is much more significant is the appearance of the two Watchers who appear to him in a dream-vision as he rests in his bed, for as the heavily reconstructed text reads:
>
>> [I saw Watchers] in my vision, the dream-vision. Two (men) were fighting over me, saying… and holding a great contest over me. I asked them, 'Who are you, that you are thus empowered over me?' They answered me, 'We] [have been empowered and rule over all mankind.' They said to me, 'Which of us do yo[u choose to rule (you)?' I raised my eyes and looked.] [One] of them was terr[i]fying in his appearance, [like a s]erpent, [his] c[loa]k many-coloured yet very dark… [And I looked again], and… in his appearance, his visage like a viper, and [wearing...] [exceedingly, and all his eyes...].
>
> The text identifies this last Watcher as Belial, the Prince of Darkness and King of Evil, while his companion is revealed as Michael, the Prince of Light, who is also named as Melchizedek, the King of Righteousness. It is, however, Belial's frightful appearance that took my attention, for he is seen as terrifying to look upon and like a 'serpent', the very synonym so often used when describing both the Watchers and the Nephilim. If the textual fragment had ended here, then I would not have known why this synonym had been used by the Jewish scribe in question. Fortunately, however, the text goes on to say that the Watcher possessed a visage, or face, 'like a viper'. Since he also wears a cloak 'many-colored yet very dark', I had also to presume that he was anthropomorphic, in other words he possessed human form.
>
> Visage like a viper...

> What could this possibly mean?
>
> How many people do you know with a 'visage like a viper'?
>
> For over a year I could offer no suitable solution to this curious metaphor.[94]

Collins says that something he heard on a national radio station gave him some sudden insight in this regard. It was about the Viper Room, the famous nightclub in Hollywood where the young actor River Phoenix was partying the night he collapsed and died of an overdose in October 1993. In the media frenzy that accompanied this event, it was noted that the Viper Room got its name years before, in the jazz era. It seems musicians would play for long hours, neglecting sleeping and eating, staying high on drugs. In the low light and smoky haze, the musicians' faces could appear hollow and gaunt, and their eyes were often closed to mere slits. Their countenance was snakelike, hence the name of the club.

Says Collins:

> This amusing anecdote sent my mind reeling and enabled me to construct a mental picture of what a person with a 'visage like a viper' might look like; their faces would appear long and narrow, with prominent cheekbones, elongated jawbones, thin lips and slanted eyes like those of many East Asian racial types. Was this the solution as to why both the Watchers and Nephilim were described as walking serpents? It seemed as likely a possibility as any, although it was also feasible that their serpentine connection related to their accredited magical associations and capabilities, perhaps even their bodily movements and overall appearance.[94]

Collins also says that the Watchers wore feathered cloaks to enhance their identity as "bird shamans" and "bird men." He claims that the Watchers would choose a bird to emulate, and then spend long periods of time studying their chosen bird's every movement. He says they would enter its natural habitat and watch every facet of its life—its method of flight, its eating habits, its courtship rituals and its actions on the ground. They did this to enhance their mental link with the birds, and create for themselves a kind of semi-permanent "alter-personality." According to Collins:

> Totemic shamanism is more-or-less dependent on the indigenous animals or birds present in the locale of the culture or tribe, although in principle the purpose has always been the same—using this mantle to achieve astral flight, divine illumination, spirit communication and the attainment of otherworldly knowledge and wisdom.

In linking the Watchers with the Anunnaki, Collins says:

> The Sumerians were a unique people with their own language and culture. Nobody knows their true origin or where exactly they may have obtained the seeds of knowledge that helped establish the various city-states during the fourth millennium BC. Yet the Sumerians themselves were quite explicit on this point. They said their entire culture had been inherited from the Anannage [Anunnaki], the gods of Anu, who had come from an ancestral homeland in the mountains. To emphasize this point they used an ideogram of a mountain to denote 'the country', ie Sumer, and built seven-tiered ziggurats in honor of these founder gods.
>
> Was it possible therefore that the proposed Watcher culture of Kurdistan provided the impetus for the rise of western civilization?
>
> Archaeologists have no problem accepting Kurdistan as the cradle of Near Eastern civilization. Shortly after the recession of the last Ice Age, c. 8500 BC, there emerged in this region some of the earliest examples of agriculture, animal domestication, baked and painted pottery, metallurgy and worked obsidian tools and utensils. Curiously enough, from c. 5750 BC onwards for several hundred years the trade in raw and worked obsidian throughout Kurdistan seems to have been centered around an extinct volcano named Nemrut Dag on the south-western shores of Lake Van, the very area in

> which both the mythical lands of Eden and Dilmun are likely to have been located.
>
> Kurdistan was undoubtedly the point of origin of the so-called Neolithic explosion from the ninth millennium BC onwards. Indeed, it is because of this settled community lifestyle in Kurdistan that the earliest known form of token bartering developed. This primitive method of exchange eventually led to the establishment of the first written alphabet and ideogram system on the Mesopotamian plains sometime during the fourth millennium BC. It is therefore understandable that civilization first arose in the Fertile Crescent during this same age. From here, of course, it quickly spread to many other regions of the Old World.[94]

So, in the end we are faced with the dual questions of whether postdiluvian civilization started in Kurdistan and whether the creators were the Watchers, also known as the Anunnaki. Zechariah Sitchin supposes that these creators were from another planet. Collins theorizes that the creators were shamanic birdmen whose people originally came from a pre-flood Egypt prior to 10,000 BC. He calls these people the Shemsu Hor, or Disciples of Horus. They practiced cranial deformation and wore feathered cloaks; with their elongated heads (and noses?) they looked like "vipers."[95]

Collins sees the Watchers as drug-using shamans who lusted after women and worshiped birds for their power of flight, they themselves seeking the ability to "astral travel." But did the Watchers actually have the ability to fly? They had the vestiges of high technology, presumably from the advanced world before the Great Flood. With the knowledge of metals comes the knowledge of forges, and the making of hard metal tools. The knowledge of electricity and flight comes shortly after the ability to create machines with hard metals.

The ancient legends of many lands tell of flying machines, sometimes called vimanas. The Watchers may have called themselves "birdmen" for more than just the ability to astral travel with the help of hallucinogenic drugs—perhaps they had the ability to actually fly with the aid of the flying machines they had built!

Now we can see the Watchers as an ancient version of "cargo cult" gods. They are humans with machines that can fly, but are seen as "gods from the sky" by the local natives who are as flesh and blood as their gods are.

Whether the Anunnaki and Watchers were angels, extraterrestrials or human leftovers from the advanced civilizations of pre-dynastic Egypt or Atlantis, we can pretty much be sure that they were coneheads. They, like the coneheads of Paracas or Tiwanaku, Malta or Mexico, had elongated heads, and in some cases, they were able to elongate their skulls in a dramatic fashion that continues to astonish us today.

There is so little material on the strange practice of cranial deformation that one has to wonder why it is such a neglected topic. As we have noted, perhaps the main reason for this is the isolationist view of history propounded by academics from universities around the world. If civilizations and cultures are largely unconnected, especially those separated by large bodies of water like the Atlantic and Pacific Oceans, then how can there be a connection between cultures that practiced cranial deformation?

Well, if oceans are barriers instead of highways, then there can't really be a connection. But since the custom of cranial deformation is so unusual and specialized, it seems unscientific to deny the obvious link between these widely separated cultures and their curious customs. For academics, it is best not to discuss the matter at all. Basically, cranial deformation as a worldwide phenomenon is a no-go area for academics, unless they want to stir up quite a bit of controversy and probably lose their tenure.

As for the fascinating concept that some skeletons with elongated skulls are of a different race from *Homo sapiens*, or that they are from another planet altogether, we have to wait for some daring and specialized DNA analysis to prove the matter one way or the other. Until then, no doubt there will be plenty of speculation; hopefully, more research will be done, and more books and articles will be written on the enigma of cranial deformation and the proliferation of coneheads in our past.

An Anunnaki god is depicted in a flying disk on this rock-cut Sumerian relief.

Three Sumerian Anunnaki are depicted inside a flying disk on this clay cylinder seal now at the British Museum.

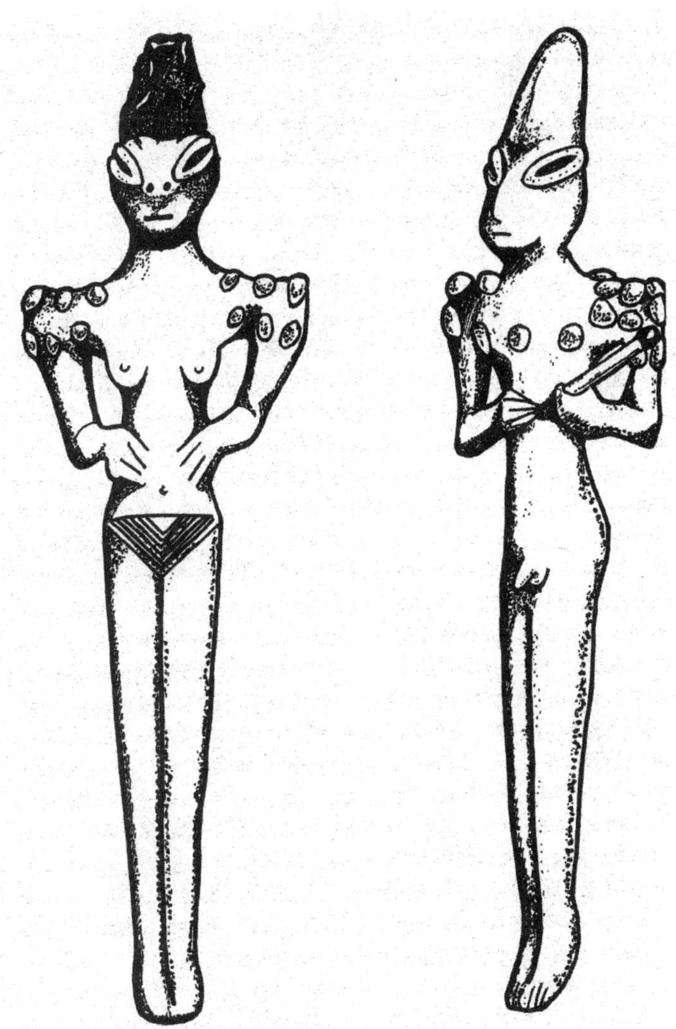

Drawing of two Sumerian clay figurines from the Ubaid Period (5000 BC) showing elongated heads and "coffeebean" eyes. The female on the left is from the ancient city of Ur and the male on the right is from Eridu.

Above and right: Terra cotta figurines from the Sumerian city of Ubaid with elongated heads and "coffeebean" eyes. These statuettes were mass produced in the region, following a uniform design.

Excavations at the Sumerian city of Ubaid now in Iraq, circa 1939.

Above, left and below: Terra cotta figurines from the Sumerian city of Ubaid. Note the extreme viper visage on the statuette above.

Left: A Sumerian stele showing humans and the Annunaki. *Below*: An Annunaki god is depicted with wings, a tall hat, a pinecone in his right hand and a mysterious satchel in his left hand.

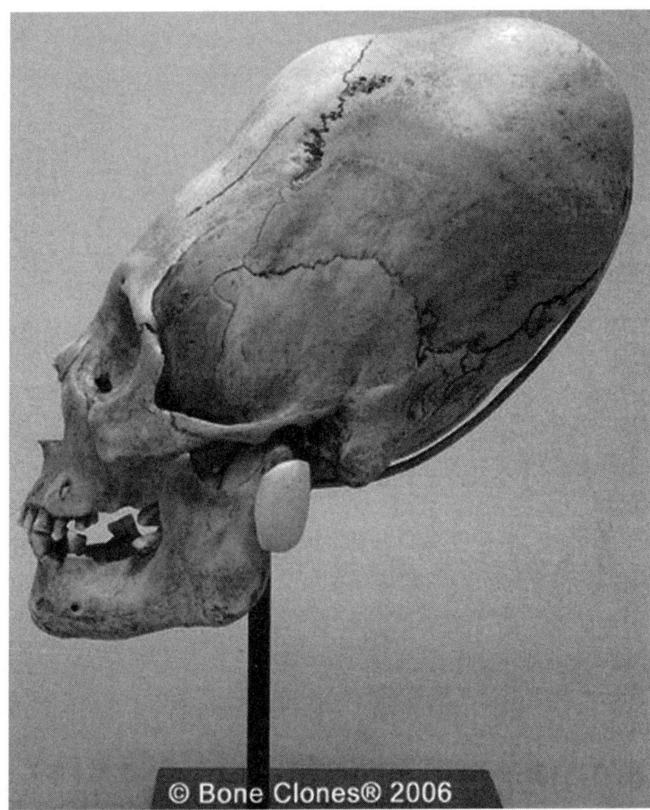

A Bone Clones reproduction of a Paracas skull for sale on their website.

A Mayan frieze, now destroyed, discovered by the archeologist Teobert Mahler in the Yucatan. Does it show the destruction of the Olmec/Mayan world over two thousand years ago?

Ojo

"A MI HIJO LO MATARON, NO MURIO EN LA CAIDA"

IMPACTO/PAG. 3

S/ 0.70 Vía aérea

DOMINGO / 20 de noviembre de 2011 S/.0.50 Director: Víctor Ramírez Canales /Año 43/ Nº 15842 www.ojo.pe

ANTROPOLOGO DICE QUE EXTRATERRESTRES "ASESORARON" A LOS INCAS

"E.T. LLEGO EN OVNI"

Doctor Dávila asegura que habrían ayudado a construir la Ciudad Imperial y Machu Picchu. Cerca a momias alienígenas fueron hallados dos meteoritos de 15 toneladas de peso

RENATO DAVILA R. (ANTROPOLOGO)

CUPÓN DE DESCUENTO

S/.5 + ESTE CUPÓN

Correo semanal ATADO DE MANOS

Explica que la unión de bloques de piedra es tecnología inca única en el planeta y que, sin haber usado cemento, rocas se unen más

The Peruvian newspaper *Ojo* had a cover story on an elongated skull discovered near Cuzco, Nov. 20, 2011. The article claimed that a Russian archeologist said it was an E.T.

Chapter 8

BIBLIOGRAPHY

1. *A History of Mexican Archaeology*, Ignacio Bernal, 1980, Ignacio Bernal, Thames & Hudson, London.

2. *Scattered Skeletons in Our Closet*, Karen Mutton, 2011, AUP, Kempton, IL.

3. *The Olmec World*, Ignacio Bernal, 1969, University of California Press, Berkeley.

4. *African Presence in Early America*, Ivan Van Sertima, 1987, Rutgers State University, New Brunswick, NJ.

5. *African Presence in Early Asia*, Ivan Van Sertima, 1987, Rutgers State University, New Brunswick, NJ.

6. *The Mystery of the Olmecs*, David Hatcher Childress, 2003, Adventures Unlimited Press, Kempton, IL.

7. *Lost Cities & Ancient Mysteries of South America*, David Hatcher Childress, 1987, AUP, Kempton, IL.

8. *Eccentric Lives and Peculiar Notions*, John Michell, 1999, Adventures Unlimited Press.

9. *1421: The Year China Discovered the World*, Gavin Menzies, 2002, Random House, New York.

10. *When China Ruled the Seas*, Louise Levathes, 1994, Simon & Schuster, New York.

11. *Olmec Archaeology and Early Mesoamerica*, Christopher Pool, 2007, Cambridge University Press, New York.

12. *The Rise and Fall of Maya Civilization*, J. Eric S. Thompson, 1954, 1966, University of Oklahoma Press, Norman, Oklahoma.

13. *Atlantis In America*, Ivar Zapp and George Erickson, 1998, AUP, Kempton, IL.

14. *Mexico*, Michael D. Coe, 1962, Thames & Hudson, London.

15. *The Discovery and Conquest of Mexico,* Bernal Diaz del Castillo, translated by A. P. Maudslay, 1956, Farrar, Straus and Cudahy, New York.

16. *Olmec Art of Ancient Mexico,* National Gallery of Art, Washington, 1996, Harry Abrams, New York.

17. *Ancient Man: A Handbook of Puzzling Artifacts*, William Corliss, 1978, The Sourcebook Project, Glen Arm, MD.

18. *Lost Cities of North and Central America*, David Hatcher Childress, 1994, Adventures Unlimited Press, Kempton, Illinois.

19. *The Olmecs: America's First Civilization,* Richard A. Diehl, 2004, Thames & Hudson, New York.

20. *In Search of Quetzalcoatl*, Pierre Honoré, 2006, Adventures Unlimited Press, Kempton, IL. (Originally *In Quest of the White God*, 1963, Hutchinson & Co. London)

21. *Los Olmecas*, Jacques Soustelle, 1979, Fondo de Cultura Economica, Mexico City.

22. *Fair Gods and Feathered Serpents*, T.J. O'Brien, 1997, Horizon Publishers, Bountiful, UT.

23. *They Came Before Columbus: The African Presence in Ancient America,* Ivan Van Sertima, 1977, Random House, New York.

24. *The Children of the Sun*, W. J. Perry, 1923, Adventures Unlimited Press, Kempton, IL.

25. *La Antropologia Americanista en la Actualidad Vol. 1*, Raphael Girard, 1980, Editores Mexicanos Unidos, Mexico City.

26. *La Antropologia Americanista en la Actualidad Vol. 2*, Raphael Girard, 1980, Editores Mexicanos Unidos, Mexico City.

27. *The Ancient Atlantic*, L. Taylor Hansen, 1969, Amherst Press, Amherst, WI.

28. *Sweat of the Sun and Tears of the Moon*, André Emmerich,1965, University of Washington Press, Seattle.

29. *Xalapa Museum of Anthropology: A Guided Tour*, 2004, Governors of the Museum, Xalapa.

30. *Chalcatzingo: Excavations on the Olmec Frontier*, David C. Grove, 1984, Thames & Hudson, New York.

31. Stringer, C.B. (1994). "Evolution of early humans". In Steve Jones, Robert Martin & David Pilbeam (eds.). *The Cambridge Encyclopedia of Human Evolution*. Cambridge: Cambridge University Press. p. 242

32. McHenry, H.M (2009). "Human Evolution". In Michael Ruse & Joseph Travis. *Evolution: The First Four Billion Years*. Cambridge, Massachusetts: The Belknap Press of Harvard University Press. p. 265

33. Stringer CB, Andrews P (March 1988). "Genetic and fossil evidence for the origin of modern humans". *Science* **239** (4845): 1263–8

34. *The American Heritage Dictionary of the English Language* (4th ed.). Houghton Mifflin Company. 2000.

35. *The Descent of Man, and Selection in Relation to Sex*, Darwin, Charles (1871. This edition published 1981, with Introduction by John Tyler Bonner & Robert M. May). Princeton, New Jersey: Princeton University Press.

36. Dart, R.A. (1953): "The Predatory Transition from Ape to Man." *International Anthropological and Linguistic Review,* 1, pp. 201–217.

37. Wood, B. & Collard, M. (1999) The changing face of Genus Homo. Evol. Anth. 8(6) 195-207.

38. Turner W (1895). "On M. Dubois' Description of Remains recently found in Java, named by him Pithecanthropus erectus: With Remarks on so-called Transitional Forms between Apes and Man". *Journal of anatomy and physiology* **29** (Pt 3): 424–45.

39. Serre D, Langaney A, Chech M, *et al*. (2004). "No Evidence of Neandertal mtDNA Contribution to Early Modern Humans". *PLoS Biol*. **2** (3): E57. doi:10.1371/journal.pbio.0020057

40. Wikipedia: *Human Evolution*.

41. Cann RL, Stoneking M, Wilson AC (1987). "Mitochondrial DNA and human evolution". *Nature* **325** (6099): 31–6. doi:10.1038/325031a0. PMID 3025745. Archived from the original on 2010-11-22.

42. Macaulay, V.; Hill, C; Achilli, A; Rengo, C; Clarke, D; Meehan, W; Blackburn, J; Semino, O et al. (2005). "Single, Rapid Coastal Settlement of Asia Revealed by Analysis of Complete Mitochondrial Genomes". *Science* **308** (5724): 1034–6. doi:10.1126/science.1109792

43. *Secret Cities of Old South America*, Harold Wilkins, 1952, reprinted by AUP, Kempton, IL.

44. *Strange World*, Frank Edwards, 1964, Bantam Books, NYC.

45. *Stranger Than Science,* Frank Edwards, 1959, Bantam Books, NYC.

46. *Nu Sun, Asian American Voyages 500 B.C.,* Gunnar Thompson, 1989, Pioneer Press, Fresno, California.

47. *The Mysterious Past*, Robert Charroux, 1973, Robert Laffont, NYC.

48. *The Conquest of New Spain*, Bernal Díaz, (1492-1580), 1963, Penguin Books, London.

49. *Enigmas*, Rupert Gould, 1945, University Books, NYC.

50. *More "Things",* Ivan T. Sanderson, 1969, Pyramid Books, NYC.

51. *The life and writings of Julio C. Tello: America's first indigenous archaeologist, pp. 1, 28 and 38-39, 72*. Richard L. Burger, 2009, University of Iowa Press.

52. Richard L. Burger, Abstract of "The Life and Writings of Julio C. Tello", University of Iowa Press, accessed 27 September 2010.

53. *Paracas Part II, Cavernas Y Necropolis*, Julio C. Tello and T. Mejia Xesspe, 1979, Universidad Nacional Mayor de San Marcos, Lima.

54. McBain, Anna, Personal conversation with the Brien Foerster, June, 2011.

55. Arnold, J. R.; Libby, W. F. (1949). "Age Determinations by Radiocarbon Content: Checks with Samples of Known Age". *Science* 110 (2869): 678–680. Bibcode 1949Sci...110..678A. doi:10.1126/science.110.2869.678

56. Navarro, Juan, Personal conversation with the Brien Foerster, July, 2011.

57. *Peruvian Antiquities,* Mariano Eduardo de Ribero y Ustáriz, Mariano Eduardo and Johann Jakob von Tschudi, A.S. Barnes & Co., London, 1854.

58. *Maps of the Ancient Sea Kings*, Charles Hapgood, 1966, Chilton Books, reprinted by Adventures Unlimited Press, Kempton, IL.

59. *People of the Jaguar*, Nicholas J. Saunders, 1989, Souvenir Press, London.

60. Denison Isa: *Der göttlicher Code* (in German), Govinda-Verlag GmbH, D-79 795 Jestetten 2004, pages 218-225, ISBN 3-906347-70-2.

61. *Men Out of Asia*, Harold Gladwin, 1947, McGraw-Hill, NYC.

62. *The Ra Expeditions,* Thor Heyerdahl, 1965, London.

63. Navarro, Juan. Personal conversation with the author October 2011.

64. Berrin, Katherine & Larco Museum. *The Spirit of Ancient Peru: Treasures from the Museo Arqueológico Rafael Larco Herrera*. New York: Thames and Hudson, 1997.

65. Wikipedia: Paracas Textiles at the British Museum

66. *A Forest of Kings*, Linda Schele & David Freidel, 1990, Morrow & Co. NYC.

67. *Prehistoric America*, Betty J. Meggers, 1972, Smithsonian Institution and Aldine Atherton, Chicago.

68. http://www.mysteriousplaces.com/Easter_Island/html/contro.html

69. Kelley, David H.; Milone, Eugene F. (November 19, 2004). *Exploring Ancient Skies: An Encyclopedic Survey of Archaeoastronomy*. Springer.

70. *Valley of the Spirits: A Journey into the Lost Realm of the Aymara*, Alan L. Kolata, John Wiley and Sons, Hoboken, NJ, 1996.

71. http://www.tenspheres.com/researches/precession.htm

72. *A Brief History of the Incas*, Brien Foerster, 2010, CMYK Impressors, Lima, Peru.

73. *Faces of the Pharaohs*, Robert B. Partridge, 1994, Rubicon Press, London.

74. *The Ancient Sun Kingdoms of the Americas*, Victor von Hagen, 1957, World Publishing Co. Cleveland, Ohio.

75. *The Ancient Maya*, Sylvanus Morley, 1946, Stanford University Press, Palo Alto, CA.

76. *Megalithic Man in America?*, George F. Carter, 1991, *The Diffusion Issue*, Stonehenge Viewpoint, Santa Barbara, California, USA.

77. *Time & Reality in the Thought of the Maya*, Miguel León-Portilla, 1988, University of Oklahoma Press, Norman, OK.

78. *Sacred Mysteries Among the Mayas & the Quiches*, Augustus LePlongeon, 1886, Kegan Paul (Agent), London & NY.

79. *Queen Moo & the Egyptian Sphinx*, Augustus LePlongeon, 1900, Kegan Paul (Agent), London & NY.

80. *Atlantic Crossings Before Columbus*, Frederick Pohl, 1961, Norton, NYC.

81. *Mystery Cities of the Maya*, Thomas Gann, 1925, Duckworth, London. Reprinted 1994, AUP, Kempton, IL.

82. *Quest For the Lost City*, Dana & Ginger Lamb, 1951, Santa Barbara Press, CA.

83. *Relación de las Cosas de Yucatan*, Friar Diego de Landa, 1579, (published as *Yucatan Before & After the Conquest,* 1937, The Maya Society, Baltimore, reissued 1978, Dover Publications, NYC).

84. *Incidents of Travel in Central America, Chiapas and Yucatan*, John L. Stevens, 1841, Harper & Bros. NY (reprinted by Dover, 1969, NYC).

85. *Saga America*, Barry Fell, 1980, New York Times Books, New York.

86. *Maya Explorer,* Victor Von Hagen, 1947, University of Oklahoma Press, Norman, OK.

87. *Mexican Cities of the Gods*, Hans Helfritz, 1968, Praeger Publishers, NYC.

88. *The Lost World of Quintana Roo*, Michel Peissel, 1963, E.P. Dutton Co. NYC.

89. *The Great Temple and the Aztec Gods*, Doris Heyden & L.F. Villaseñor, 1984, Editorial Minutiae Mexicana, Mexico City.

90. *Oaxaca: The Archaeological Record*, Marcus Winter, 1989, Editorial Minutiae Mexicana, Mexico City.

91. *The Oaxaca Valley*, 1973, Instituto Nacional de Antropología e Historia, Mexico City.

92. *Mexico: From the Olmecs to the Aztecs*, Michael D. Coe, 2002;

Thames and Hudson, London.

93. *Hindu America?*, Chaman Lal, 1960, Bharatiya Vidya Bhavan, Bombay.

94. *From the Ashes of Angels*, Andrew Collins, 1996, Penguin Books, London.

95. *Gods of Eden*, Andrew Collins, 1998, Hodder-Headline, London.

96. *They All Discovered America*, Charles Boland, 1961, Doubleday, Garden City, New York.

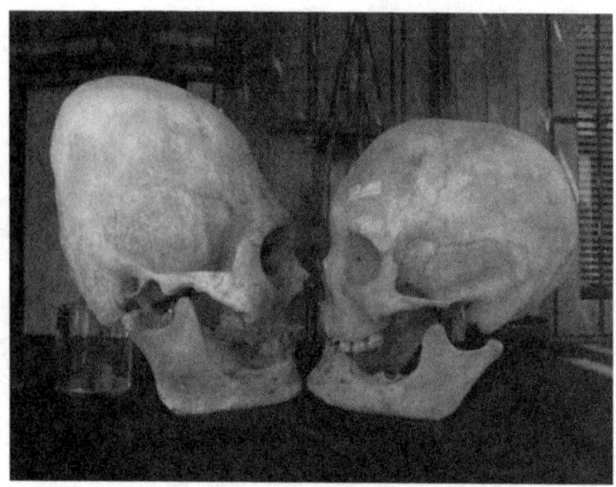

Two of the skulls at the Paracas History Museum.

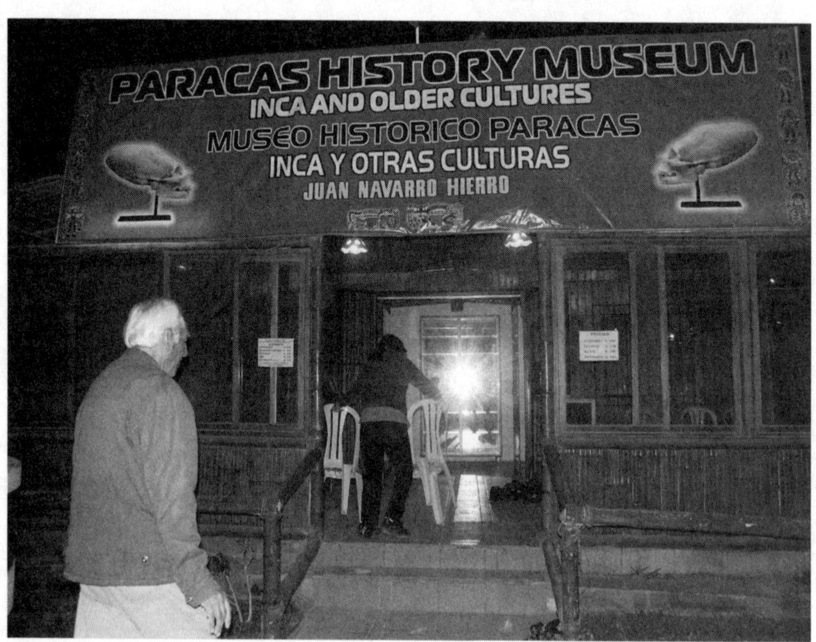

The entrance to Paracas History Museum where some elongated skulls can be seen.

Get these fascinating books from your nearest bookstore or directly from: Adventures Unlimited Press

www.adventuresunlimitedpress.com

SECRETS OF THE MYSTERIOUS VALLEY

by Christopher O'Brien

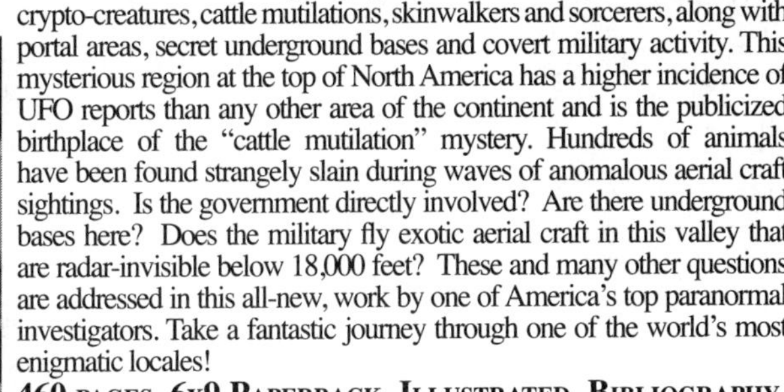

No other region in North America features the variety and intensity of unusual phenomena found in the world's largest alpine valley, the San Luis Valley of Colorado and New Mexico. Since 1989, Christopher O'Brien has documented thousands of high-strange accounts that report UFOs, ghosts, crypto-creatures, cattle mutilations, skinwalkers and sorcerers, along with portal areas, secret underground bases and covert military activity. This mysterious region at the top of North America has a higher incidence of UFO reports than any other area of the continent and is the publicized birthplace of the "cattle mutilation" mystery. Hundreds of animals have been found strangely slain during waves of anomalous aerial craft sightings. Is the government directly involved? Are there underground bases here? Does the military fly exotic aerial craft in this valley that are radar-invisible below 18,000 feet? These and many other questions are addressed in this all-new, work by one of America's top paranormal investigators. Take a fantastic journey through one of the world's most enigmatic locales!

460 PAGES. 6X9 PAPERBACK. ILLUSTRATED. BIBLIOGRAPHY. $19.95. CODE: SOMV

LOST CITIES OF NORTH & CENTRAL AMERICA

by David Hatcher Childress

Down the back roads from coast to coast, maverick archaeologist and adventurer David Hatcher Childress goes deep into unknown America. With this incredible book, you will search for lost Mayan cities and books of gold, discover an ancient canal system in Arizona, climb gigantic pyramids in the Midwest, explore megalithic monuments in New England, and join the astonishing quest for lost cities throughout North America. From the war-torn jungles of Guatemala, Nicaragua and Honduras to the deserts, mountains and fields of Mexico, Canada, and the U.S.A., Childress takes the reader in search of sunken ruins, Viking forts, strange tunnel systems, living dinosaurs, early Chinese explorers, and fantastic lost treasure. Packed with both early and current maps, photos and illustrations.

590 PAGES. 6X9 PAPERBACK. ILLUSTRATED. FOOTNOTES. BIBLIOGRAPHY. INDEX. $16.95. CODE: NCA

LOST CITIES & ANCIENT MYSTERIES OF SOUTH AMERICA

David Hatcher Childress

LOST CITIES & ANCIENT MYSTERIES OF SOUTH AMERICA

by David Hatcher Childress

Rogue adventurer and maverick archaeologist David Hatcher Childress takes the reader on unforgettable journeys deep into deadly jungles, high up on windswept mountains and across scorching deserts in search of lost civilizations and ancient mysteries. Travel with David and explore stone cities high in mountain forests and hear fantastic tales of Inca treasure, living dinosaurs, and a mysterious tunnel system. Whether he is hopping freight trains, searching for secret cities, or just dealing with the daily problems of food, money, and romance, the author keeps the reader spellbound. Includes both early and current maps, photos, and illustrations, and plenty of advice for the explorer planning his or her own journey of discovery.

381 PAGES. 6X9 PAPERBACK. ILLUSTRATED. FOOTNOTES. BIBLIOGRAPHY. INDEX. $16.95. CODE: SAM

LOST CITIES & ANCIENT MYSTERIES OF AFRICA & ARABIA

by David Hatcher Childress

Childress continues his world-wide quest for lost cities and ancient mysteries. Join him as he discovers forbidden cities in the Empty Quarter of Arabia; "Atlantean" ruins in Egypt and the Kalahari desert; a mysterious, ancient empire in the Sahara; and more. This is the tale of an extraordinary life on the road: across war-torn countries, Childress searches for King Solomon's Mines, living dinosaurs, the Ark of the Covenant and the solutions to some of the fantastic mysteries of the past.

423 PAGES. 6X9 PAPERBACK. ILLUSTRATED. $14.95. CODE: AFA

LOST CITIES OF ATLANTIS, ANCIENT EUROPE & THE MEDITERRANEAN

by David Hatcher Childress

Childress takes the reader in search of sunken cities in the Mediterranean; across the Atlas Mountains in search of Atlantean ruins; to remote islands in search of megalithic ruins; to meet living legends and secret societies. From Ireland to Turkey, Morocco to Eastern Europe, and around the remote islands of the Mediterranean and Atlantic, Childress takes the reader on an astonishing quest for mankind's past. Ancient technology, cataclysms, megalithic construction, lost civilizations and devastating wars of the past are all explored in this book.

524 PAGES. 6X9 PAPERBACK. ILLUSTRATED. $16.95. CODE: MED

LOST CITIES OF CHINA, CENTRAL ASIA & INDIA

by David Hatcher Childress

Like a real life "Indiana Jones," maverick archaeologist David Childress takes the reader on an incredible adventure across some of the world's oldest and most remote countries in search of lost cities and ancient mysteries. Discover ancient cities in the Gobi Desert; hear fantastic tales of lost continents, vanished civilizations and secret societies bent on ruling the world; visit forgotten monasteries in forbidding snow-capped mountains with strange tunnels to mysterious subterranean cities! A unique combination of far-out exploration and practical travel advice, it will astound and delight the experienced traveler or the armchair voyager.

429 PAGES. 6X9 PAPERBACK. ILLUSTRATED. FOOTNOTES & BIBLIOGRAPHY. $14.95. CODE: CHI

LOST CITIES OF ANCIENT LEMURIA & THE PACIFIC

by David Hatcher Childress

Was there once a continent in the Pacific? Called Lemuria or Pacifica by geologists, Mu or Pan by the mystics, there is now ample mythological, geological and archaeological evidence to "prove" that an advanced and ancient civilization once lived in the central Pacific. Maverick archaeologist and explorer David Hatcher Childress combs the Indian Ocean, Australia and the Pacific in search of the surprising truth about mankind's past. Contains photos of the underwater city on Pohnpei; explanations on how the statues were levitated around Easter Island in a clockwise vortex movement; tales of disappearing islands; Egyptians in Australia; and more.

379 PAGES. 6X9 PAPERBACK. ILLUSTRATED. FOOTNOTES & BIBLIOGRAPHY. $14.95. CODE: LEM

SUNKEN REALMS

A Survey of Underwater Ruins Around the World

By Karen Mutton

Australian researcher Karen Mutton begins by discussing some of the causes for sunken ruins: super-floods; volcanoes; earthquakes at the end of the last great flood; plate tectonics and other theories. From there she launches into a worldwide cataloging of underwater ruins by region. She begins with the many underwater cities in the Mediterranean, and then moves into northern Europe and the North Atlantic. Places covered in this book include: Tartessos; Cadiz; Morocco; Alexandria; The Bay of Naples; Libya; Phoenician and Egyptian sites; Roman era sites; Yarmuta, Lebanon; Cyprus; Malta; Thule & Hyperborea; Canary and Azore Islands; Bahamas; Cuba; Bermuda; Peru; Micronesia; Japan; Indian Ocean; Sri Lanka Land Bridge; Lake Titicaca; and inland lakes in Scotland, Russia, Iran, China, Wisconsin, Florida and more.

320 Pages. 6x9 Paperback. Illustrated. Bibliography. $20.00. Code: SRLM

TECHNOLOGY OF THE GODS

The Incredible Sciences of the Ancients

by David Hatcher Childress

Childress looks at the technology that was allegedly used in Atlantis and the theory that the Great Pyramid of Egypt was originally a gigantic power station. He examines tales of ancient flight and the technology that it involved; how the ancients used electricity; megalithic building techniques; the use of crystal lenses and the fire from the gods; evidence of various high tech weapons in the past, including atomic weapons; ancient metallurgy and heavy machinery; the role of modern inventors such as Nikola Tesla in bringing ancient technology back into modern use; impossible artifacts; and more.

356 PAGES. 6x9 PAPERBACK. ILLUSTRATED. BIBLIOGRAPHY. $16.95. CODE: TGOD

VIMANA AIRCRAFT OF ANCIENT INDIA & ATLANTIS

by David Hatcher Childress, introduction by Ivan T. Sanderson

In this incredible volume on ancient India, authentic Indian texts such as the *Ramayana* and the *Mahabharata* are used to prove that ancient aircraft were in use more than four thousand years ago. Included in this book is the entire Fourth Century BC manuscript *Vimaanika Shastra* by the ancient author Maharishi Bharadwaaja. Also included are chapters on Atlantean technology, the incredible Rama Empire of India and the devastating wars that destroyed it.

334 PAGES. 6x9 PAPERBACK. ILLUSTRATED. $15.95. CODE: VAA

LOST CONTINENTS & THE HOLLOW EARTH

I Remember Lemuria and the Shaver Mystery

by David Hatcher Childress & Richard Shaver

Shaver's rare 1948 book *I Remember Lemuria* is reprinted in its entirety, and the book is packed with illustrations from Ray Palmer's *Amazing Stories* magazine of the 1940s. Palmer and Shaver told of tunnels running through the earth—tunnels inhabited by the Deros and Teros, humanoids from an ancient spacefaring race that had inhabited the earth, eventually going underground, hundreds of thousands of years ago. Childress discusses the famous hollow earth books and delves deep into whatever reality may be behind the stories of tunnels in the earth. Operation High Jump to Antarctica in 1947 and Admiral Byrd's bizarre statements, tunnel systems in South America and Tibet, the underground world of Agartha, the belief of UFOs coming from the South Pole, more.

344 PAGES. 6x9 PAPERBACK. ILLUSTRATED. $16.95. CODE: LCHE

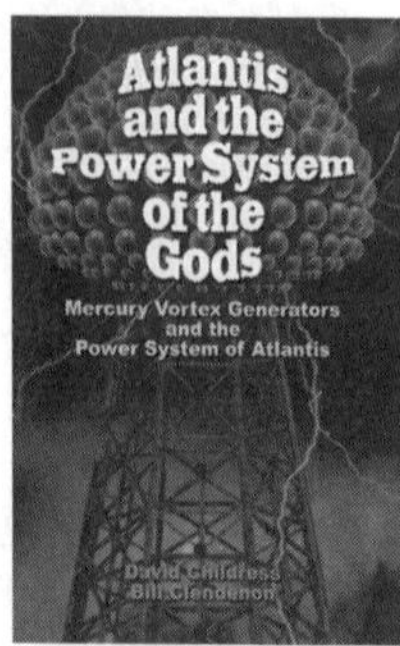

ATLANTIS & THE POWER SYSTEM OF THE GODS

by David Hatcher Childress and Bill Clendenon

Childress' fascinating analysis of Nikola Tesla's broadcast system in light of Edgar Cayce's "Terrible Crystal" and the obelisks of ancient Egypt and Ethiopia. Includes: Atlantis and its crystal power towers that broadcast energy; how these incredible power stations may still exist today; inventor Nikola Tesla's nearly identical system of power transmission; Mercury Proton Gyros and mercury vortex propulsion; more. Richly illustrated, and packed with evidence that Atlantis not only existed—it had a world-wide energy system more sophisticated than ours today.

246 PAGES. 6X9 PAPERBACK. ILLUSTRATED. $15.95. CODE: APSG

THE ANTI-GRAVITY HANDBOOK

edited by David Hatcher Childress

The new expanded compilation of material on Anti-Gravity, Free Energy, Flying Saucer Propulsion, UFOs, Suppressed Technology, NASA Cover-ups and more. Highly illustrated with patents, technical illustrations and photos. This revised and expanded edition has more material, including photos of Area 51, Nevada, the government's secret testing facility. This classic on weird science is back in a new format!

230 PAGES. 7X10 PAPERBACK. ILLUSTRATED. $16.95. CODE: AGH

ANTI–GRAVITY & THE WORLD GRID

Is the earth surrounded by an intricate electromagnetic grid network offering free energy? This compilation of material on ley lines and world power points contains chapters on the geography, mathematics, and light harmonics of the earth grid. Learn the purpose of ley lines and ancient megalithic structures located on the grid. Discover how the grid made the Philadelphia Experiment possible. Explore the Coral Castle and many other mysteries, including acoustic levitation, Tesla Shields and scalar wave weaponry. Browse through the section on anti-gravity patents, and research resources.

274 PAGES. 7X10 PAPERBACK. ILLUSTRATED. $14.95. CODE: AGW

ANTI–GRAVITY & THE UNIFIED FIELD

edited by David Hatcher Childress

Is Einstein's Unified Field Theory the answer to all of our energy problems? Explored in this compilation of material is how gravity, electricity and magnetism manifest from a unified field around us. Why artificial gravity is possible; secrets of UFO propulsion; free energy; Nikola Tesla and anti-gravity airships of the 20s and 30s; flying saucers as superconducting whirls of plasma; anti-mass generators; vortex propulsion; suppressed technology; government cover-ups; gravitational pulse drive; spacecraft & more.

240 PAGES. 7X10 PAPERBACK. ILLUSTRATED. $14.95. CODE: AGU

THE TIME TRAVEL HANDBOOK

A Manual of Practical Teleportation & Time Travel

edited by David Hatcher Childress

The Time Travel Handbook takes the reader beyond the government experiments and deep into the uncharted territory of early time travellers such as Nikola Tesla and Guglielmo Marconi and their alleged time travel experiments, as well as the Wilson Brothers of EMI and their connection to the Philadelphia Experiment—the U.S. Navy's forays into invisibility, time travel, and teleportation. Childress looks into the claims of time travelling individuals, and investigates the unusual claim that the pyramids on Mars were built in the future and sent back in time. A highly visual, large format book, with patents, photos and schematics. Be the first on your block to build your own time travel device!

316 PAGES. 7X10 PAPERBACK. ILLUSTRATED. $16.95. CODE: TTH

MAPS OF THE ANCIENT SEA KINGS
Evidence of Advanced Civilization in the Ice Age
by Charles H. Hapgood

Charles Hapgood has found the evidence in the Piri Reis Map that shows Antarctica, the Hadji Ahmed map, the Oronteus Finaeus and other amazing maps. Hapgood concluded that these maps were made from more ancient maps from the various ancient archives around the world, now lost. Not only were these unknown people more advanced in mapmaking than any people prior to the 18th century, it appears they mapped all the continents. The Americas were mapped thousands of years before Columbus. Antarctica was mapped when its coasts were free of ice!

316 PAGES. 7X10 PAPERBACK. ILLUSTRATED. BIBLIOGRAPHY & INDEX. $19.95. CODE: MASK

PATH OF THE POLE
Cataclysmic Pole Shift Geology
by Charles H. Hapgood

Maps of the Ancient Sea Kings author Hapgood's classic book *Path of the Pole* is back in print! Hapgood researched Antarctica, ancient maps and the geological record to conclude that the Earth's crust has slipped on the inner core many times in the past, changing the position of the pole. *Path of the Pole* discusses the various "pole shifts" in Earth's past, giving evidence for each one, and moves on to possible future pole shifts.

356 PAGES. 6X9 PAPERBACK. ILLUSTRATED. $16.95. CODE: POP

SECRETS OF THE HOLY LANCE
The Spear of Destiny in History & Legend
by Jerry E. Smith

Secrets of the Holy Lance traces the Spear from its possession by Constantine, Rome's first Christian Caesar, to Charlemagne's claim that with it he ruled the Holy Roman Empire by Divine Right, and on through two thousand years of kings and emperors, until it came within Hitler's grasp—and beyond! Did it rest for a while in Antarctic ice? Is it now hidden in Europe, awaiting the next person to claim its awesome power? Neither debunking nor worshiping, *Secrets of the Holy Lance* seeks to pierce the veil of myth and mystery around the Spear. Mere belief that it was infused with magic by virtue of its shedding the Savior's blood has made men kings. But what if it's more? What are "the powers it serves"?

312 PAGES. 6X9 PAPERBACK. ILLUSTRATED. BIBLIOGRAPHY. $16.95. CODE: SOHL

THE FANTASTIC INVENTIONS OF NIKOLA TESLA
by Nikola Tesla with additional material by David Hatcher Childress

This book is a readable compendium of patents, diagrams, photos and explanations of the many incredible inventions of the originator of the modern era of electrification. In Tesla's own words are such topics as wireless transmission of power, death rays, and radio-controlled airships. In addition, rare material on a secret city built at a remote jungle site in South America by one of Tesla's students, Guglielmo Marconi. Marconi's secret group claims to have built flying saucers in the 1940s and to have gone to Mars in the early 1950s! Incredible photos of these Tesla craft are included. •His plan to transmit free electricity into the atmosphere. •How electrical devices would work using only small antennas. •Why unlimited power could be utilized anywhere on earth. •How radio and radar technology can be used as death-ray weapons in Star Wars.

342 PAGES. 6X9 PAPERBACK. ILLUSTRATED. $16.95. CODE: FINT

REICH OF THE BLACK SUN

Nazi Secret Weapons & the Cold War Allied Legend

by Joseph P. Farrell

Why were the Allies worried about an atom bomb attack by the Germans in 1944? Why did the Soviets threaten to use poison gas against the Germans? Why did Hitler in 1945 insist that holding Prague could win the war for the Third Reich? Why did US General George Patton's Third Army race for the Skoda works at Pilsen in Czechoslovakia instead of Berlin? Why did the US Army not test the uranium atom bomb it dropped on Hiroshima? Why did the Luftwaffe fly a non-stop round trip mission to within twenty miles of New York City in 1944? *Reich of the Black Sun* takes the reader on a scientific-historical journey in order to answer these questions. Arguing that Nazi Germany actually won the race for the atom bomb in late 1944,

352 PAGES. 6x9 PAPERBACK. ILLUSTRATED. BIBLIOGRAPHY. $16.95. CODE: ROBS

THE GIZA DEATH STAR

The Paleophysics of the Great Pyramid & the Military Complex at Giza

by Joseph P. Farrell

Was the Giza complex part of a military installation over 10,000 years ago? Chapters include: An Archaeology of Mass Destruction, Thoth and Theories; The Machine Hypothesis; Pythagoras, Plato, Planck, and the Pyramid; The Weapon Hypothesis; Encoded Harmonics of the Planck Units in the Great Pyramid; High Fregquency Direct Current "Impulse" Technology; The Grand Gallery and its Crystals: Gravito-acoustic Resonators; The Other Two Large Pyramids; the "Causeways," and the "Temples"; A Phase Conjugate Howitzer; Evidence of the Use of Weapons of Mass Destruction in Ancient Times; more.

290 PAGES. 6x9 PAPERBACK. ILLUSTRATED. $16.95. CODE: GDS

THE GIZA DEATH STAR DEPLOYED

The Physics & Engineering of the Great Pyramid

by Joseph P. Farrell

Farrell expands on his thesis that the Great Pyramid was a maser, designed as a weapon and eventually deployed—with disastrous results to the solar system. Includes: Exploding Planets: A Brief History of the Exoteric and Esoteric Investigations of the Great Pyramid; No Machines, Please!; The Stargate Conspiracy; The Scalar Weapons; Message or Machine?; A Tesla Analysis of the Putative Physics and Engineering of the Giza Death Star; Cohering the Zero Point, Vacuum Energy, Flux: Feedback Loops and Tetrahedral Physics; and more.

290 PAGES. 6x9 PAPERBACK. ILLUSTRATED. $16.95. CODE: GDSD

THE GIZA DEATH STAR DESTROYED

The Ancient War For Future Science

by Joseph P. Farrell

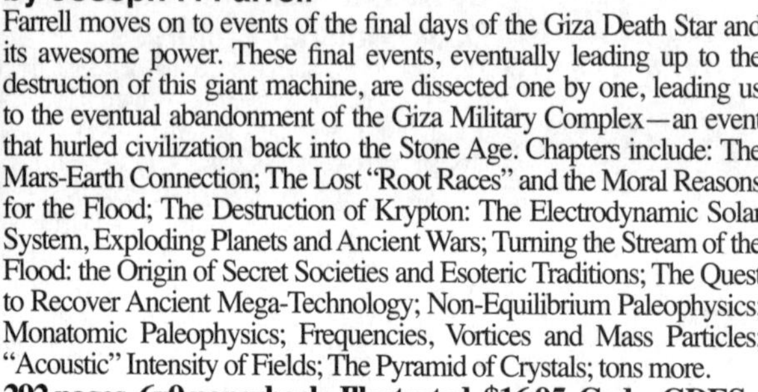

Farrell moves on to events of the final days of the Giza Death Star and its awesome power. These final events, eventually leading up to the destruction of this giant machine, are dissected one by one, leading us to the eventual abandonment of the Giza Military Complex—an event that hurled civilization back into the Stone Age. Chapters include: The Mars-Earth Connection; The Lost "Root Races" and the Moral Reasons for the Flood; The Destruction of Krypton: The Electrodynamic Solar System, Exploding Planets and Ancient Wars; Turning the Stream of the Flood: the Origin of Secret Societies and Esoteric Traditions; The Quest to Recover Ancient Mega-Technology; Non-Equilibrium Paleophysics; Monatomic Paleophysics; Frequencies, Vortices and Mass Particles; "Acoustic" Intensity of Fields; The Pyramid of Crystals; tons more.

292 pages. 6x9 paperback. Illustrated. $16.95. Code: GDES

THE TESLA PAPERS

Nikola Tesla on Free Energy & Wireless Transmission of Power

by Nikola Tesla, edited by David Hatcher Childress

David Hatcher Childress takes us into the incredible world of Nikola Tesla and his amazing inventions. Tesla's fantastic vision of the future, including wireless power, anti-gravity, free energy and highly advanced solar power. Also included are some of the papers, patents and material collected on Tesla at the Colorado Springs Tesla Symposiums, including papers on: •The Secret History of Wireless Transmission •Tesla and the Magnifying Transmitter •Design and Construction of a Half-Wave Tesla Coil •Electrostatics: A Key to Free Energy •Progress in Zero-Point Energy Research •Electromagnetic Energy from Antennas to Atoms •Tesla's Particle Beam Technology •Fundamental Excitatory Modes of the Earth-Ionosphere Cavity

325 PAGES. 8X10 PAPERBACK. ILLUSTRATED. $16.95. CODE: TTP

UFOS AND ANTI-GRAVITY

Piece For A Jig-Saw

by Leonard G. Cramp

Leonard G. Cramp's 1966 classic book on flying saucer propulsion and suppressed technology is a highly technical look at the UFO phenomena by a trained scientist. Cramp first introduces the idea of 'anti-gravity' and introduces us to the various theories of gravitation. He then examines the technology necessary to build a flying saucer and examines in great detail the technical aspects of such a craft. Cramp's book is a wealth of material and diagrams on flying saucers, anti-gravity, suppressed technology, G-fields and UFOs. Chapters include Crossroads of Aerodymanics, Aerodynamic Saucers, Limitations of Rocketry, Gravitation and the Ether, Gravitational Spaceships, G-Field Lift Effects, The Bi-Field Theory, VTOL and Hovercraft, Analysis of UFO photos, more.

388 PAGES. 6X9 PAPERBACK. ILLUSTRATED. $16.95. CODE: UAG

THE COSMIC MATRIX

Piece for a Jig-Saw, Part Two

by Leonard G. Cramp

Cramp examines anti-gravity effects and theorizes that this super-science used by the craft—described in detail in the book—can lift mankind into a new level of technology, transportation and understanding of the universe. The book takes a close look at gravity control, time travel, and the interlocking web of energy between all planets in our solar system with Leonard's unique technical diagrams. A fantastic voyage into the present and future!

364 PAGES. 6X9 PAPERBACK. ILLUSTRATED. BIBLIOGRAPHY. $16.00. CODE: CMX

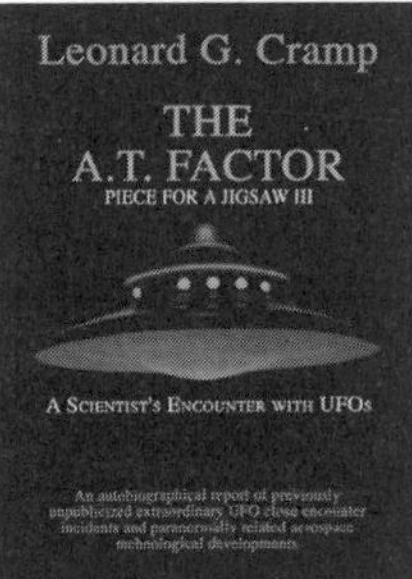

THE A.T. FACTOR

A Scientists Encounter with UFOs

by Leonard Cramp

British aerospace engineer Cramp began much of the scientific anti-gravity and UFO propulsion analysis back in 1955 with his landmark book *Space, Gravity & the Flying Saucer* (out-of-print and rare). In this final book, Cramp brings to a close his detailed and controversial study of UFOs and Anti-Gravity.

324 PAGES. 6X9 PAPERBACK. ILLUSTRATED. BIBLIOGRAPHY. INDEX. $16.95. CODE: ATF

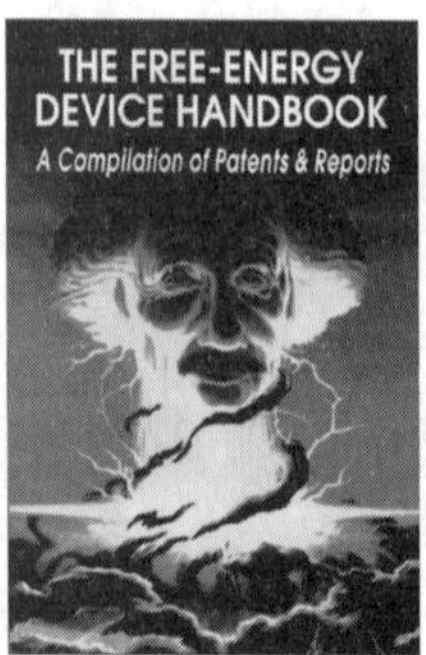

THE FREE-ENERGY DEVICE HANDBOOK
A Compilation of Patents and Reports
by David Hatcher Childress

A large-format compilation of various patents, papers, descriptions and diagrams concerning free-energy devices and systems. *The Free-Energy Device Handbook* is a visual tool for experimenters and researchers into magnetic motors and other "over-unity" devices. With chapters on the Adams Motor, the Hans Coler Generator, cold fusion, superconductors, "N" machines, space-energy generators, Nikola Tesla, T. Townsend Brown, and the latest in free-energy devices. Packed with photos, technical diagrams, patents and fascinating information, this book belongs on every science shelf.

292 PAGES. 8X10 PAPERBACK. ILLUSTRATED. $16.95. CODE: FEH

THE ENERGY GRID
Harmonic 695, The Pulse of the Universe
by Captain Bruce Cathie

This is the breakthrough book that explores the incredible potential of the Energy Grid and the Earth's Unified Field all around us. Cathie's first book, *Harmonic 33*, was published in 1968 when he was a commercial pilot in New Zealand. Since then, Captain Bruce Cathie has been the premier investigator into the amazing potential of the infinite energy that surrounds our planet every microsecond. Cathie investigates the Harmonics of Light and how the Energy Grid is created. In this amazing book are chapters on UFO Propulsion, Nikola Tesla, Unified Equations, the Mysterious Aerials, Pythagoras & the Grid, Nuclear Detonation and the Grid, Maps of the Ancients, an Australian Stonehenge examined, more.

255 PAGES. 6X9 TRADEPAPER. ILLUSTRATED. $15.95. CODE: TEG

THE BRIDGE TO INFINITY
Harmonic 371244
by Captain Bruce Cathie

Cathie has popularized the concept that the earth is crisscrossed by an electromagnetic grid system that can be used for anti-gravity, free energy, levitation and more. The book includes a new analysis of the harmonic nature of reality, acoustic levitation, pyramid power, harmonic receiver towers and UFO propulsion. It concludes that today's scientists have at their command a fantastic store of knowledge with which to advance the welfare of the human race.

204 PAGES. 6X9 TRADEPAPER. ILLUSTRATED. $14.95. CODE: BTF

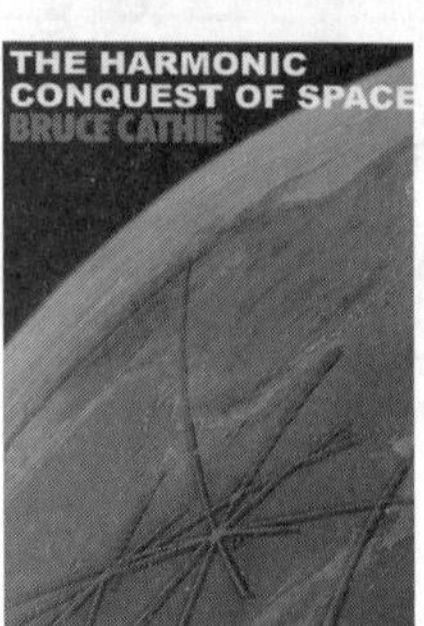

THE HARMONIC CONQUEST OF SPACE
by Captain Bruce Cathie

Chapters include: Mathematics of the World Grid; the Harmonics of Hiroshima and Nagasaki; Harmonic Transmission and Receiving; the Link Between Human Brain Waves; the Cavity Resonance between the Earth; the Ionosphere and Gravity; Edgar Cayce—the Harmonics of the Subconscious; Stonehenge; the Harmonics of the Moon; the Pyramids of Mars; Nikola Tesla's Electric Car; the Robert Adams Pulsed Electric Motor Generator; Harmonic Clues to the Unified Field; and more. Also included are tables showing the harmonic relations between the earth's magnetic field, the speed of light, and anti-gravity/gravity acceleration at different points on the earth's surface. New chapters in this edition on the giant stone spheres of Costa Rica, Atomic Tests and Volcanic Activity, and a chapter on Ayers Rock analysed with Stone Mountain, Georgia.

248 PAGES. 6X9. PAPERBACK. ILLUSTRATED. BIBLIOGRAPHY. $16.95. CODE: HCS

GRAVITATIONAL MANIPULATION OF DOMED CRAFT
UFO Propulsion Dynamics
by Paul E. Potter

Potter's precise and lavish illustrations allow the reader to enter directly into the realm of the advanced technological engineer and to understand, quite straightforwardly, the aliens' methods of energy manipulation: their methods of electrical power generation; how they purposely designed their craft to employ the kinds of energy dynamics that are exclusive to space (discoverable in our astrophysics) in order that their craft may generate both attractive and repulsive gravitational forces; their control over the mass-density matrix surrounding their craft enabling them to alter their physical dimensions and even manufacture their own frame of reference in respect to time. Includes a 16-page color insert.

624 pages. 7x10 Paperback. Illustrated. References. $24.00. Code: GMDC

TAPPING THE ZERO POINT ENERGY
Free Energy & Anti-Gravity in Today's Physics
by Moray B. King

King explains how free energy and anti-gravity are possible. The theories of the zero point energy maintain there are tremendous fluctuations of electrical field energy imbedded within the fabric of space. This book tells how, in the 1930s, inventor T. Henry Moray could produce a fifty kilowatt "free energy" machine; how an electrified plasma vortex creates anti-gravity; how the Pons/Fleischmann "cold fusion" experiment could produce tremendous heat without fusion; and how certain experiments might produce a gravitational anomaly.

180 PAGES. 5x8 PAPERBACK. ILLUSTRATED. $12.95. CODE: TAP

QUEST FOR ZERO-POINT ENERGY
Engineering Principles for "Free Energy"
by Moray B. King

King expands, with diagrams, on how free energy and anti-gravity are possible. The theories of zero point energy maintain there are tremendous fluctuations of electrical field energy embedded within the fabric of space. King explains the following topics: TFundamentals of a Zero-Point Energy Technology; Vacuum Energy Vortices; The Super Tube; Charge Clusters: The Basis of Zero-Point Energy Inventions; Vortex Filaments, Torsion Fields and the Zero-Point Energy; Transforming the Planet with a Zero-Point Energy Experiment; Dual Vortex Forms: The Key to a Large Zero-Point Energy Coherence. Packed with diagrams, patents and photos.

224 PAGES. 6x9 PAPERBACK. ILLUSTRATED. $14.95. CODE: QZPE

DARK MOON
Apollo and the Whistleblowers
by Mary Bennett and David Percy

Did you know a second craft was going to the Moon at the same time as Apollo 11? Do you know that potentially lethal radiation is prevalent throughout deep space? Do you know there are serious discrepancies in the account of the Apollo 13 'accident'? Did you know that 'live' color TV from the Moon was not actually live at all? Did you know that the Lunar Surface Camera had no viewfinder? Do you know that lighting was used in the Apollo photographs—yet no lighting equipment was taken to the Moon? All these questions, and more, are discussed in great detail by British researchers Bennett and Percy in *Dark Moon*, the definitive book (nearly 600 pages) on the possible faking of the Apollo Moon missions. Tons of NASA photos analyzed for possible deceptions.

568 PAGES. 6x9 PAPERBACK. ILLUSTRATED. BIBLIOGRAPHY. INDEX. $32.00. CODE: DMO

THE MYSTERY OF THE OLMECS
by David Hatcher Childress

The Olmecs were not acknowledged to have existed as a civilization until an international archeological meeting in Mexico City in 1942. Now, the Olmecs are slowly being recognized as the Mother Culture of Mesoamerica, having invented writing, the ball game and the "Mayan" Calendar. But who were the Olmecs? Where did they come from? What happened to them? How sophisticated was their culture? Why are many Olmec statues and figurines seemingly of foreign peoples such as Africans, Europeans and Chinese? Is there a link with Atlantis? In this heavily illustrated book, join Childress in search of the lost cities of the Olmecs! Chapters include: The Mystery of Quizuo; The Mystery of Transoceanic Trade; The Mystery of Cranial Deformation; more.

296 PAGES. 6x9 PAPERBACK. ILLUSTRATED. BIBLIOGRAPHY. COLOR SECTION. $20.00. CODE: MOLM

THE LAND OF OSIRIS
An Introduction to Khemitology
by Stephen S. Mehler

Was there an advanced prehistoric civilization in ancient Egypt who built the great pyramids and carved the Great Sphinx? Did the pyramids serve as energy devices and not as tombs for kings? Mehler has uncovered an indigenous oral tradition that still exists in Egypt, and has been fortunate to have studied with a living master of this tradition, Abd'El Hakim Awyan. Mehler has also been given permission to present these teachings to the Western world, teachings that unfold a whole new understanding of ancient Egypt . Chapters include: Egyptology and Its Paradigms; Asgat Nefer—The Harmony of Water; Khemit and the Myth of Atlantis; The Extraterrestrial Question; more.

272 PAGES. 6x9 PAPERBACK. ILLUSTRATED. COLOR SECTION. BIBLIOGRAPHY. $18.00 CODE: LOOS

ABOMINABLE SNOWMEN: LEGEND COME TO LIFE
The Story of Sub-Humans on Six Continents from the Early Ice Age Until Today
by Ivan T. Sanderson

Do "Abominable Snowmen" exist? Prepare yourself for a shock. In the opinion of one of the world's leading naturalists, not one, but possibly four kinds, still walk the earth! Do they really live on the fringes of the towering Himalayas and the edge of myth-haunted Tibet? From how many areas in the world have factual reports of wild, strange, hairy men emanated? Reports of strange apemen have come in from every continent, except Antarctica.

525 PAGES. 6x9 PAPERBACK. ILLUSTRATED. BIBLIOGRAPHY. INDEX. $16.95. CODE: ABML

INVISIBLE RESIDENTS
The Reality of Underwater UFOS
by Ivan T. Sanderson

In this book, Sanderson, a renowned zoologist with a keen interest in the paranormal, puts forward the curious theory that "OINTS"—Other Intelligences—live under the Earth's oceans. This underwater, parallel, civilization may be twice as old as Homo sapiens, he proposes, and may have "developed what we call space flight." Sanderson postulates that the OINTS are behind many UFO sightings as well as the mysterious disappearances of aircraft and ships in the Bermuda Triangle. What better place to have an impenetrable base than deep within the oceans of the planet? Sanderson offers here an exhaustive study of USOs (Unidentified Submarine Objects) observed in nearly every part of the world.

298 PAGES. 6x9 PAPERBACK. ILLUSTRATED. BIBLIOGRAPHY. INDEX. $16.95. CODE: INVS

PIRATES & THE LOST TEMPLAR FLEET
The Secret Naval War Between the Templars & the Vatican
by David Hatcher Childress

Childress takes us into the fascinating world of maverick sea captains who were Knights Templar (and later Scottish Rite Free Masons) who battled the ships that sailed for the Pope. The lost Templar fleet was originally based at La Rochelle in southern France, but fled to the deep fiords of Scotland upon the dissolution of the Order by King Phillip. This banned fleet of ships was later commanded by the St. Clair family of Rosslyn Chapel (birthplace of Free Masonry). St. Clair and his Templars made a voyage to Canada in the year 1298 AD, nearly 100 years before Columbus! Later, this fleet of ships and new ones to come, flew the Skull and Crossbones, the symbol of the Knights Templar.

320 PAGES. 6x9 PAPERBACK. ILLUSTRATED. BIBLIOGRAPHY. $16.95. CODE: PLTF

TEMPLARS' LEGACY IN MONTREAL
The New Jerusalem
by Francine Bernier

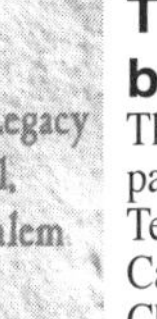

The book reveals the links between Montreal and: John the Baptist as patron saint; Melchizedek, the first king-priest and a father figure to the Templars and the Essenes; Stella Maris, the Star of the Sea from Mount Carmel; the Phrygian goddess Cybele as the androgynous Mother of the Church; St. Blaise, the Armenian healer or "Therapeut"- the patron saint of the stonemasons and a major figure to the Benedictine Order and the Templars; the presence of two Black Virgins; an intriguing family coat of arms with twelve blue apples; and more.

352 PAGES. 6x9 PAPERBACK. ILLUSTRATED. BIBLIOGRAPHY. $21.95. CODE: TLIM

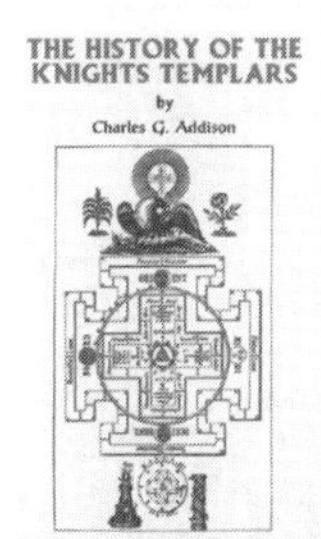

THE HISTORY OF THE KNIGHTS TEMPLARS
by Charles G. Addison, introduction by David Hatcher Childress

Chapters on the origin of the Templars, their popularity in Europe and their rivalry with the Knights of St. John, later to be known as the Knights of Malta. Detailed information on the activities of the Templars in the Holy Land, and the 1312 AD suppression of the Templars in France and other countries, which culminated in the execution of Jacques de Molay and the continuation of the Knights Templars in England and Scotland; the formation of the society of Knights Templars in London; and the rebuilding of the Temple in 1816. Plus a lengthy intro about the lost Templar fleet and its North American sea routes.

395 PAGES. 6x9 PAPERBACK. ILLUSTRATED. $16.95. CODE: HKT

OTTO RAHN & THE QUEST FOR THE HOLY GRAIL
The Amazing Life of the Real "Indiana Jones"
by Nigel Graddon

Otto Rahn led a life of incredible adventure in southern France in the early 1930s. The Hessian language scholar is said to have found runic Grail tablets in the Pyrenean grottoes, and decoded hidden messages within the medieval Grail masterwork *Parsifal*. The fabulous artifacts identified by Rahn were believed by Himmler to include the Grail Cup, the Spear of Destiny, the Tablets of Moses, the Ark of the Covenant, the Sword and Harp of David, the Sacred Candelabra and the Golden Urn of Manna. Some believe that Rahn was a Nazi guru who wielded immense influence on his elders and "betters" within the Hitler regime, persuading them that the Grail was the Sacred Book of the Aryans, which, once obtained, would justify their extreme political theories and revivify the ancient Germanic myths. But things are never as they seem, and as new facts emerge about Otto Rahn a far more extraordinary story unfolds.

450 pages. 6x9 Paperback. Illustrated. $18.95. Code: ORQG

CASEBOOK ON THE MEN IN BLACK

By Jim Keith, Foreword by Kenn Thomas

UFO witnesses are sometimes intimidated by mysterious men dressed entirely in black. Are they government agents, sinister aliens or interdimensional creatures? Keith chronicles the strange goings on surrounding UFO activity and often bizarre cars that they arrive in—literal flying cars! Chapters include: Black Arts; Demons and Witches; Black Lodge; Maury Island; On a Bender; The Silence Group; Overlords and UMMO; More Black Ops; Indrid Cold; M.I.B.s in a Test Tube; Green Yard; The Hoaxers; Gray Areas; You Will Cease UFO Study; Beyond Reality; The Real/Unreal Men in Black; Deciphering a Nightmare; more. A new edition of the classic with updated material!

222 pages. 5x9 Paperback. Illustrated. $14.95. Code: CMIB

EYE OF THE PHOENIX

Mysterious Visions and Secrets of the American Southwest by Gary David

GaryDavid explores enigmas and anomalies in the vast American Southwest. Contents includes: The Great Pyramids of Arizona; Meteor Crater—Arizona's First Bonanza?; Chaco Canyon—Ancient City of the Dog Star; Phoenix—Masonic Metropolis in the Valley of the Sun; Along the 33rd Parallel—A Global Mystery Circle; The Flying Shields of the Hopi Katsinam; Is the Starchild a Hopi God?; The Ant People of Orion—Ancient Star Beings of the Hopi; Serpent Knights of the Round Temple; The Nagas—Origin of the Hopi Snake Clan?; The Tau (or T-shaped) Cross—Hopi/Maya/Egyptian Connections; The Hopi Stone Tablets of Techqua Ikachi; The Four Arms of Destiny—Swastikas in the Hopi World of the End Times; and more.

348 pages. 6x9 Paperback. Illustrated. $16.95. Code: EOPX

PRODIGAL GENIUS

The Life of Nikola Tesla by John J. O'Neill

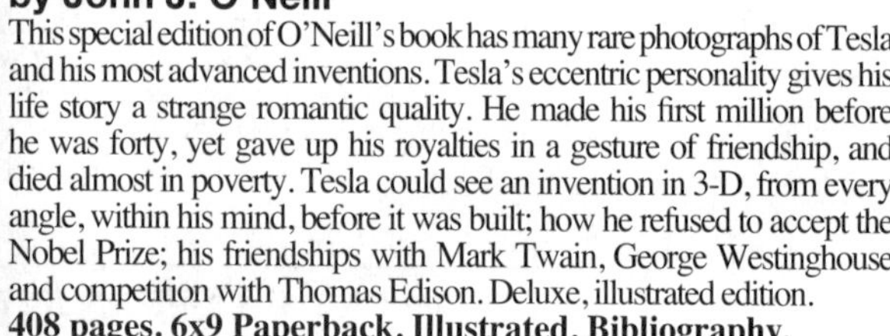

This special edition of O'Neill's book has many rare photographs of Tesla and his most advanced inventions. Tesla's eccentric personality gives his life story a strange romantic quality. He made his first million before he was forty, yet gave up his royalties in a gesture of friendship, and died almost in poverty. Tesla could see an invention in 3-D, from every angle, within his mind, before it was built; how he refused to accept the Nobel Prize; his friendships with Mark Twain, George Westinghouse and competition with Thomas Edison. Deluxe, illustrated edition.

408 pages. 6x9 Paperback. Illustrated. Bibliography. $18.95. Code: PRG

STALKING THE TRICKSTERS:

Shapeshifters, Skinwalkers, Dark Adepts and 2012 By Christopher O'Brien Foreword by David Perkins

Manifestations of the Trickster persona such as cryptids, elementals, werewolves, demons, vampires and dancing devils have permeated human experience since before the dawn of civilization. But today, very little is publicly known about The Tricksters. Who are they? What is their agenda? Known by many names including fools, sages, Loki, men-in-black, skinwalkers, shapeshifters, jokers, *jinn,* sorcerers, and witches, Tricksters provide us with a direct conduit to the unknown in the 21st century. Can these denizens of phenomenal events be attempting to communicate a warning to humanity in this uncertain age of prophesied change? Take a journey around the world stalking the tricksters!

354 Pages. 6x9 Paperback. Illustrated. Bibliography. $18.95. Code: STT

THE CRYSTAL SKULLS
Astonishing Portals to Man's Past
by David Hatcher Childress and Stephen S. Mehler

Childress introduces the technology and lore of crystals, and then plunges into the turbulent times of the Mexican Revolution form the backdrop for the rollicking adventures of Ambrose Bierce, the renowned journalist who went missing in the jungles in 1913, and F.A. Mitchell-Hedges, the notorious adventurer who emerged from the jungles with the most famous of the crystal skulls. Mehler shares his extensive knowledge of and experience with crystal skulls. Having been involved in the field since the 1980s, he has personally examined many of the most influential skulls, and has worked with the leaders in crystal skull research, including the inimitable Nick Nocerino, who developed a meticulous methodology for the purpose of examining the skulls.

294 pages. 6x9 Paperback. Illustrated. Bibliography. $18.95. Code: CRSK

THE INCREDIBLE LIGHT BEINGS OF THE COSMOS
Are Orbs Intelligent Light Beings from the Cosmos?
by Antonia Scott-Clark

Scott-Clark has experienced orbs for many years, but started photographing them in earnest in the year 2000 when the "Light Beings" entered her life. She took these very seriously and set about privately researching orb occurrences. The incredible results of her findings are presented here, along with many of her spectacular photographs. With her friend, GoGos lead singer Belinda Carlisle, Antonia tells of her many adventures with orbs. Find the answers to questions such as: Can you see orbs with the naked eye?; Are orbs intelligent?; What are the Black Villages?; What is the connection between orbs and crop circles? Antonia gives detailed instruction on how to photograph orbs, and how to communicate with these Light Beings of the Cosmos.

334 pages. 6x9 Paperback. Illustrated. References. $19.95. Code: ILBC

AXIS OF THE WORLD
The Search for the Oldest American Civilization
by Igor Witkowski

Polish author Witkowski's research reveals remnants of a high civilization that was able to exert its influence on almost the entire planet, and did so with full consciousness. Sites around South America show that this was not just one of the places influenced by this culture, but a place where they built their crowning achievements. Easter Island, in the southeastern Pacific, constitutes one of them. The Rongo-Rongo language that developed there points westward to the Indus Valley. Taken together, the facts presented by Witkowski provide a fresh, new proof that an antediluvian, great civilization flourished several millennia ago.

220 pages. 6x9 Paperback. Illustrated. References. $18.95. Code: AXOW

LEY LINE & EARTH ENERGIES
An Extraordinary Journey into the Earth's Natural Energy System
by David Cowan & Chris Arnold

The mysterious standing stones, burial grounds and stone circles that lace Europe, the British Isles and other areas have intrigued scientists, writers, artists and travellers through the centuries. How do ley lines work? How did our ancestors use Earth energy to map their sacred sites and burial grounds? How do ghosts and poltergeists interact with Earth energy? How can Earth spirals and black spots affect our health? This exploration shows how natural forces affect our behavior, how they can be used to enhance our health and well being.

368 pages. 6x9 Paperback. Illustrated. $18.95. Code: LLEE

SECRETS OF THE UNIFIED FIELD
The Philadelphia Experiment, the Nazi Bell, and the Discarded Theory
by Joseph P. Farrell

Farrell examines the now discarded Unified Field Theory. American and German wartime scientists and engineers determined that, while the theory was incomplete, it could nevertheless be engineered. Chapters include: The Meanings of "Torsion"; Wringing an Aluminum Can; The Mistake in Unified Field Theories and Their Discarding by Contemporary Physics; Three Routes to the Doomsday Weapon: Quantum Potential, Torsion, and Vortices; Tesla's Meeting with FDR; Arnold Sommerfeld and Electromagnetic Radar Stealth; Electromagnetic Phase Conjugations, Phase Conjugate Mirrors, and Templates; The Unified Field Theory, the Torsion Tensor, and Igor Witkowski's Idea of the Plasma Focus; tons more.

340 pages. 6x9 Paperback. Illustrated. Bibliography. Index. $18.95. Code: SOUF

NAZI INTERNATIONAL
The Nazi's Postwar Plan to Control Finance, Conflict, Physics and Space
by Joseph P. Farrell

Beginning with prewar corporate partnerships in the USA he moves on to the surrender of Nazi Germany, and evacuation plans of the Germans. He then covers the vast, and still-little-known recreation of Nazi Germany in South America with help of Juan Peron, I.G. Farben and Martin Bormann. Farrell then covers Nazi Germany's penetration of the Muslim world before moving on to the development and control of new energy technologies including the Bariloche Fusion Project, Dr. Philo Farnsworth's Plasmator, and the work of Dr. Nikolai Kozyrev. Finally, Farrell discusses the Nazi desire to control space, and examines their connection with NASA, the esoteric meaning of NASA Mission Patches.

412 pages. 6x9 Paperback. Illustrated. References. $19.95. Code: NZIN

ARKTOS
The Myth of the Pole in Science, Symbolism, and Nazi Survival
by Joscelyn Godwin

A scholarly treatment of catastrophes, ancient myths and the Nazi Occult beliefs. Explored are the many tales of an ancient race said to have lived in the Arctic regions, such as Thule and Hyperborea. Progressing onward, the book looks at modern polar legends including the survival of Hitler, German bases in Antarctica, UFOs, the hollow earth, Agartha and Shambala, more.

220 pages. 6x9 Paperback. Illustrated. $16.95. Code: ARK

GUARDIANS OF THE HOLY GRAIL
by Mark Amaru Pinkham

This book presents this extremely ancient Holy Grail lineage from Asia and how the Knights Templar were initiated into it. It also reveals how the ancient Asian wisdom regarding the Holy Grail became the foundation for the Holy Grail legends of the west while also serving as the bedrock of the European Secret Societies, which included the Freemasons, Rosicrucians, and the Illuminati. Also: The Fisher Kings; The Middle Eastern mystery schools, such as the Assassins and Yezidhi; The ancient Holy Grail lineage from Sri Lanka and the Templar Knights' initiation into it; The head of John the Baptist and its importance to the Templars; The secret Templar initiation with grotesque Baphomet, the infamous Head of Wisdom; more.

248 pages. 6x9 Paperback. Illustrated. $16.95. Code: GOHG

SCATTERED SKELETONS IN OUR CLOSET

By Karen Mutton

Australian researcher Mutton gives us the rundown on various hominids, skeletons, anomalous skulls and other "things" from our family tree, including hobbits, pygmies, giants and horned people. Chapters include: Human Origin Theories; Dating Techniques; Mechanisms of Darwinian Evolution; What Creationists Believe about Human Origins; Evolution Fakes and Mistakes; Creationist Hoaxes and Mistakes; The Tangled Tree of Evolution; The Australopithecine Debate; Homo Habilis; Homo Erectus; Anatomically Modern Humans in Ancient Strata?; Ancient Races of the Americas; Robust Australian Prehistoric Races; Pre Maori Races of New Zealand; The Taklamakan Mummies—Caucasians in Prehistoric China; Strange Skulls; Dolichocephaloids (Coneheads); Pumpkin Head, M Head, Horned Skulls; The Adena Skull; The Boskop Skulls; 'Starchild'; Pygmies of Ancient America; Pedro the Mountain Mummy; Hobbits—Homo Floresiensis; Palau Pygmies; Giants; Goliath; Holocaust of American Giants?; Giants from Around the World; more. Heavily illustrated.

320 Pages. 6x9 Paperback. Illustrated. $18.95. Code: SSIC

THE GRID OF THE GODS

The Aftermath of the Cosmic War and the Physics of the Pyramid Peoples

By Joseph P. Farrell with Scott D. de Hart

Physicist and Oxford-educated historian Farrell continues his best-selling book series on ancient planetary warfare, technology and the energy grid that surrounds the earth. Chapters on: Anomalies at the Temples of Angkor; The Ancient Prime Meridian: Giza; Transmitters, Temples, Sacred Sites and Nazis; Nazis and Geomancy; Nazi Transmitters and the Earth Grid; The Grid and Hitler's East Prussia Headquarters; Grid Geopolitical Geomancy; The Astronomical Correlation and the 10,500 BC Mystery; The Master Plan of a Hidden Elite; Moving and Immoveable Stones; Uncountable Stones and Stones of the Giants and Gods; Gateway Traditions; The Grid and the Ancient Elite; Finding the Center of the Land; The Ancient Catastrophe, the Very High Civilization, and the Post-Catastrophe Elite; The Meso- and South-American "Pyramid Peoples"; Tiahuanaco and the Puma Punkhu Paradox: Ancient Machining; The Mayans, Their Myths and the Mounds; The Pythagorean and Platonic Principles of Sumer, Babylonia and Greece; The Gears of Giza: the Center of the Machine; Alchemical Cosmology and Quantum Mechanics in Stone: The Mysterious Megalith of Nabta Playa; The Physics of the "Pyramid Peoples"; tons more.

436 Pages. 6x9 Paperback. Illustrated. References. $19.95. Code: GOG

BEYOND EINSTEIN'S UNIFIED FIELD

Gravity and Electro-Magnetism Redefined

By John Brandenburg, Ph.D.

Veteran plasma physicist John Brandenburg shows the intricate interweaving of Einstein's work with that of other physicists, including Sarkharov and his "zero point" theory of gravity and the hidden fifth dimension of Kaluza and Klein. He also traces the surprising, hidden influence of Nikola Tesla on Einstein's life. Brandenburg describes control of space-time geometry through electromagnetism, and states that faster-than-light travel will be possible in the future. Anti-gravity through electromagnetism is possible, which upholds the basic "flying saucer" design utilizing "The Tesla Vortex." See the physics used at Area 51 explained! Chapters include: Squaring the Circle, Einstein's Final Triumph; Mars Hill, or the Cosmos As It Is; A Book of Numbers and Forms; Kepler, Newton and the Sun King; Magnus and Electra; Atoms of Light; Einstein's Glory, Relativity; The Aurora; Tesla's Vortex and the Cliffs of Zeno; The Hidden 5th Dimension; The GEM Unification Theory; Anti-Gravity and Human Flight; The New GEM Cosmos; Summit of Mount Einstein; more. Includes and 8-page color section.

312 Pages. 6x9 Paperback. Illustrated. References. $18.95. Code: BEUF

THE BOOK OF ENOCH
translated by Richard Laurence

This is a reprint of the Apocryphal *Book of Enoch the Prophet* which was first discovered in Abyssinia in the year 1773 by a Scottish explorer named James Bruce. One of the main influences from the book is its explanation of evil coming into the world with the arrival of the "fallen angels." Enoch acts as a scribe, writing up a petition on behalf of these fallen angels, or fallen ones, to be given to a higher power for ultimate judgment. Christianity adopted some ideas from Enoch, including the Final Judgment, the concept of demons, the origins of evil and the fallen angels, and the coming of a Messiah and ultimately, a Messianic kingdom.

224 PAGES. 6X9 PAPERBACK. ILLUSTRATED. INDEX. $16.95. CODE: BOE

SUNS OF GOD
Krishna, Buddha and Christ Unveiled
by Acharya S

Over the past several centuries, the Big Three spiritual leaders have been the Lords Christ, Krishna and Buddha, whose stories and teachings are so remarkably similar as to confound and amaze those who encounter them. As classically educated archaeologist, historian, mythologist and linguist Acharya S thoroughly reveals, these striking parallels exist not because these godmen were "historical" personages who "walked the earth" but because they are personifications of the central focus of the famous and scandalous "mysteries." These mysteries date back thousands of years and are found globally, reflecting an ancient tradition steeped in awe and intrigue.

428 PAGES. 6X9 PAPERBACK. ILLUSTRATED. BIBLIOGRAPHY. INDEX. $18.95. CODE: SUNG

THE CHRIST CONSPIRACY
The Greatest Story Ever Sold
by Acharya S.

In this highly controversial and explosive book, archaeologist, historian, mythologist and linguist Acharya S. marshals an enormous amount of startling evidence to demonstrate that Christianity and the story of Jesus Christ were created by members of various secret societies, mystery schools and religions in order to unify the Roman Empire under one state religion. In developing such a fabrication, this multinational cabal drew upon a multitude of myths and rituals that existed long before the Christian era, and reworked them for centuries into the religion passed down to us today. Contrary to popular belief, Jesus was many characters rolled into one. These characters personified the ubiquitous solar myth, and their exploits were well known, as reflected by such popular deities as Mithras, Heracles/Hercules, Dionysos and many others throughout the Roman Empire and beyond.

436 PAGES. 6X9 PAPERBACK. ILLUSTRATED. $16.95. CODE: CHRC

EDEN IN EGYPT
by Ralph Ellis

The story of Adam and Eve from the Book of Genesis is perhaps one of the best-known stories in circulation, even today, and yet nobody really knows where this tale came from or what it means. But even a cursory glance at the text will demonstrate the origins of this tale, for the river of Eden is described as having four branches. There is only one river in this part of the world that fits this description, and that is the Nile, with the four branches forming the Nile Delta. According to Ellis, Judaism was based upon the reign of the pharaoh Akhenaton, because the solitary Judaic god was known as Adhon while this pharaoh's solitary god was called Aton or Adjon. But what of the identities of Adam and Eve? Includes 16 page color section.

320 PAGES. 6X9 PAPERBACK. ILLUSTRATED. BIBLIOGRAPHY. INDEX. $20.00. CODE: EIE

THE ORION ZONE
Ancient Star Cities of the American Southwest
by Gary A. David

This book on ancient star lore explores the mysterious location of Pueblos in the American Southwest, circa 1100 AD, that appear to be a mirror image of the major stars of the Orion constellation. Packed with maps, diagrams, astronomical charts, and photos of ruins and rock art, *The Orion Zone* explores this terrestrial-celestial relationship and its astounding global significance. Chapters include: Leaving Many Footprints—The Emergence and Migrations of the Anazazi; The Sky Over the Hopi Villages; Orion Rising in the Dark Crystal; The Cosmo-Magical Cities of the Anazazi; Windows Onto the Cosmos; To Calibrate the March of Time; They Came from Across the Ocean—The Patki (Water) Clan and the Snake Clan of the Hopi; Ancient and Mysterious Monuments; Beyond That Fiery Day; more.

346 pages. 6x9 Paperback. Illustrated. $19.95. Code: OZON

THE MYSTERY OF THE OLMECS
by David Hatcher Childress

The Olmecs were not acknowledged to have existed as a civilization until an international archeological meeting in Mexico City in 1942. Now, the Olmecs are slowly being recognized as the Mother Culture of Mesoamerica, having invented writing, the ball game and the "Mayan" Calendar. But who were the Olmecs? Where did they come from? What happened to them? How sophisticated was their culture? Why are many Olmec statues and figurines seemingly of foreign peoples such as Africans, Europeans and Chinese? Is there a link with Atlantis? In this heavily illustrated book, join Childress in search of the lost cities of the Olmecs! Chapters include: The Mystery of Quizuo; The Mystery of Transoceanic Trade; The Mystery of Cranial Deformation; more.

296 Pages. 6x9 Paperback. Illustrated. Bibliography. Color Section. $20.00. Code: MOLM

THE LAND OF OSIRIS
An Introduction to Khemitology
by Stephen S. Mehler

Was there an advanced prehistoric civilization in ancient Egypt who built the great pyramids and carved the Great Sphinx? Did the pyramids serve as energy devices and not as tombs for kings? Mehler has uncovered an indigenous oral tradition that still exists in Egypt, and has been fortunate to have studied with a living master of this tradition, Abd'El Hakim Awyan. Mehler has also been given permission to present these teachings to the Western world, teachings that unfold a whole new understanding of ancient Egypt . Chapters include: Egyptology and Its Paradigms; Asgat Nefer—The Harmony of Water; Khemit and the Myth of Atlantis; The Extraterrestrial Question; more.

272 pages. 6x9 Paperback. Illustrated. Color Section. Bibliography. $18.00 Code: LOOS

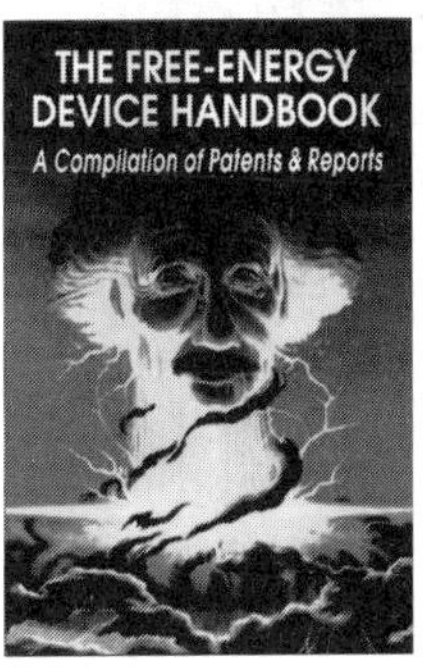

THE FREE-ENERGY DEVICE HANDBOOK
A Compilation of Patents and Reports
by David Hatcher Childress

A large-format compilation of various patents, papers, descriptions and diagrams concerning free-energy devices and systems. *The Free-Energy Device Handbook* is a visual tool for experimenters and researchers into magnetic motors and other "over-unity" devices. With chapters on the Adams Motor, the Hans Coler Generator, cold fusion, superconductors, "N" machines, space-energy generators, Nikola Tesla, T. Townsend Brown, and the latest in free-energy devices. Packed with photos, technical diagrams, patents and fascinating information, this book belongs on every science shelf.

292 pages. 8x10 Paperback. Illustrated. $16.95. Code: FEH

ORDER FORM

10% Discount When You Order 3 or More Items!

One Adventure Place
P.O. Box 74
Kempton, Illinois 60946
United States of America
Tel.: 815-253-6390 • Fax: 815-253-6300
Email: auphq@frontiernet.net
http://www.adventuresunlimitedpress.com

ORDERING INSTRUCTIONS

- ✓ Remit by USD$ Check, Money Order or Credit Card
- ✓ Visa, Master Card, Discover & AmEx Accepted
- ✓ Paypal Payments Can Be Made To: info@wexclub.com
- ✓ Prices May Change Without Notice
- ✓ 10% Discount for 3 or more Items

SHIPPING CHARGES

United States

- ✓ Postal Book Rate { $4.00 First Item / 50¢ Each Additional Item
- ✓ POSTAL BOOK RATE Cannot Be Tracked!
- ✓ Priority Mail { $5.00 First Item / $2.00 Each Additional Item
- ✓ UPS { $6.00 First Item / $1.50 Each Additional Item

NOTE: UPS Delivery Available to Mainland USA Only

Canada

- ✓ Postal Air Mail { $10.00 First Item / $2.50 Each Additional Item
- ✓ Personal Checks or Bank Drafts MUST BE US$ and Drawn on a US Bank
- ✓ Canadian Postal Money Orders OK
- ✓ Payment MUST BE US$

All Other Countries

- ✓ Sorry, No Surface Delivery!
- ✓ Postal Air Mail { $16.00 First Item / $6.00 Each Additional Item
- ✓ Checks and Money Orders MUST BE US$ and Drawn on a US Bank or branch.
- ✓ Paypal Payments Can Be Made in US$ To: info@wexclub.com

SPECIAL NOTES

- ✓ RETAILERS: Standard Discounts Available
- ✓ BACKORDERS: We Backorder all Out-of-Stock Items Unless Otherwise Requested
- ✓ PRO FORMA INVOICES: Available on Request

ORDER ONLINE AT: www.adventuresunlimitedpress.com

Please check: ☑

☐ This is my first order ☐ I have ordered before

Name
Address
City
State/Province Postal Code
Country
Phone day Evening
Fax Email

Item Code	Item Description	Qty	Total

Please check: ☑

☐ Postal-Surface
☐ Postal-Air Mail (Priority in USA)
☐ UPS (Mainland USA only)

Subtotal ▸
Less Discount-10% for 3 or more items ▸
Balance ▸
Illinois Residents 6.25% Sales Tax ▸
Previous Credit ▸
Shipping ▸
Total (check/MO in USD$ only) ▸

☐ Visa/MasterCard/Discover/American Express

Card Number

Expiration Date

10% Discount When You Order 3 or More Items!